Greenhouse
Earth

Greenhouse Earth

363.7392
N 1

Annika Nilsson

*Published on behalf of
the Scientific Committee on Problems of the Environment (SCOPE)
of the International Council of Scientific Unions (ICSU) and
the United Nations Environment Programme (UNEP)*

by

JOHN WILEY & SONS

Chichester · New York · Brisbane · Toronto · Singapore

SCOPE—Scientific Committee on Problems of the Environment,
51 boulevard de Montmorency,
75016 Paris, France
Telex 645554 F ICSU

UNEP—United Nations Environment Programme,
PO Box 30552,
Nairobi, Kenya
Telex 22068 UNEP KE

Library of Congress Cataloging-in-Publication Data

Nilsson, Annika.
    Greenhouse earth / Annika Nilsson.
        p.   cm.
        ISBN 0–471–93628–6 (paper)
    1. Greenhouse effect. Atmospheric.   2. Environmental protection.
    I. International Council of Scientific Unions. Scientific Committee
    on Problems of the Environment.   II. United Nations Environment
    Programme.   III. Title.
    QC912.3.N55   1992
    574.5′222—dc20   92–8327
                                                      CIP

British Library Cataloguing in Publication Data

A catalogue record for this book is available from the British Library

ISBN 0-471-93628-6

Typeset in 11/13pt Baskerville from author's disks by
Dobbie Typesetting Ltd, Tavistock, Devon
Printed and bound in Great Britain by
Biddles Ltd, Guildford and King's Lynn

Contents

Preface   ix

Acknowledgements   xiii

CHAPTER 1   THE HISTORY OF OUR FUTURE . . .   1
            An emerging science   6
            Laying the foundation   8

CHAPTER 2   THE GREENHOUSE EARTH   11
            Today's climate   15

CHAPTER 3   THE GREENHOUSE GASES   21
            Carbon dioxide (CO_2)   23
            The tracks of imbalance   24
            Emission of carbon dioxide   27
            Deforestation   29
            Methane (CH_4)   32
            A levelling increase   36
            Nitrous oxide (N_2O)   37
            CFCs and other halocarbons   40
            Stratospheric ozone   43
            Tropospheric ozone   44
            Aerosols   45

CHAPTER 4   ADDING UP THE NUMBERS   49
            Energy budgets   51
            Overlapping windows   53
            Concentration and time   53
            A focus on emissions   54
            Solar input   57
            Changing orbits   58

CHAPTER 5   A MODEL WORLD   61
            Icy reflections   64
            Heights and clouds   64
            Taking in all the dimensions   65
            Global confidence and regional doubt   67
            Oceans and clouds   68
            A system in transition   71
            Validating models   72
            Drier or wetter soil?   74
            Extreme weather   75
            Looking at the past   76
            A conclusion of confidence   78

CHAPTER 6   MODEL RESULTS — THE FUTURE
            CLIMATE   79
            Higher temperatures   81
            Speeding up the water cycle   83
            Soil moisture   84
            How can the sea and ice change
            the climate?   85
            A more extreme climate?   86
            Stormy patterns   87
            Regional changes in climate   88

CHAPTER 7   SEA LEVEL   91
            Measuring changes in the past   93
            Greenland and Antarctica   94
            An Antarctic collapse?   96

A higher sea 98
Regional effects 100
Threatened wetlands 103
The fishing industry 104

CHAPTER 8   AGRICULTURE 107
Carbon dioxide as a fertilizer 111
Carbon dioxide for water conservation 113
A combination of effects 114
Climate change and a shift in risks 115
A focus on the crops 116
Soils and pests 119
A summary by region 122
Global food security 129
Regional economy 133
Adjusting to change 135
Cutting emissions 136
Agriculture and deforestation 138
Cutting excess nitrogen 139
Carbon in soil 139

CHAPTER 9   FORESTRY 141
Mapping vegetation types 143
The time of death and renewal . . . 144
Pests, pollution and soil quality . . . 147
A higher output? 151
Summarizing impacts 153
Problem or possibility? 154
Reforestation 157

CHAPTER 10   NATURAL ECOSYSTEMS 159
Who will cope and who will not? . 162
One stress added to another 164

CHAPTER 11   WATER, SNOW AND ICE 167
Sensitive riversheds 169

	Who will have a white Christmas?	171
	On frozen grounds..............	174
CHAPTER 12	ANSWERING THE CHALLENGE ...	177
	International politics	179
	Energy potentials to tap	181
	The process has started	187
	Scenarios of change	191
	Energy demand and economic growth	193
	Wait and see?	198
CHAPTER 13	A GAME OF SCIENCE AND POLITICS	203
AFTERWORD	A LOOK BACK..................	213
FURTHER READING		217

Preface

C LIMATE CHANGE HAS BECOME ONE OF THE major issues on the international environmental agenda. Predictions of a rising sea and devastating droughts have alerted politicians worldwide to the risks of continued increases in the emission of carbon dioxide and other greenhouse gases.

But to change the direction of development is not an easy process. A myriad of political decisions have to be made on a national as well as international level. Those decisions need to be based on facts. The question is then how big a problem climate change really is. How much do the scientists know about what is in store?

Since the greenhouse effect and climate warming were first brought up on the international agenda of environmental problems, many efforts have been made to critically evaluate the scientific base for the predictions about climate change. The first thorough review was a joint effort by three international organizations: the International Council of Scientific Unions (ICSU), the United Nations Environment Programme (UNEP), and the World Meteorological Organization (WMO). It was published by the Scientific Committee on Problems of the Environment (SCOPE) in the report *The Greenhouse Effect, Climatic Change and Ecosystems* (SCOPE 29). This was the work of a large number of scientists who critically reviewed the existing information on emission of greenhouse gases and their fate in the atmosphere. They looked at climate models and studied past changes in climate trying to assess the potential for sea-level change and the effects of climate and carbon dioxide on agriculture and forestry.

In 1988, WMO and UNEP established the Inter-governmental Panel on Climate Change (IPCC). Two years later, this panel presented an international consensus report based on work and discussions by more than a hundred scientists. It contains a scientific assessment of the state of the atmosphere and potential for changes, an assessment of the impact of climate change and a review of policy options on how to respond to the predicted changes.

These works have provided a scientific base for the discussions about a global climate convention. They are not light reading, however, and most policy makers have to depend on short summaries of the conclusions. But when science enters the political arena and conclusions are challenged, short summaries may not provide enough information. We would all like to understand and form an opinion of our own. We like to be able to weigh the evidence with our own scale.

The United Nations Environment Programme and the Scientific Committee on Problems of the Environment therefore asked science writer Annika Nilsson to write a book about the greenhouse effect and climate change aimed at a non-scientific audience. Professor Bo R. Döös and Dr. John S. Perry assisted in scientific review of the manuscript.

This book is an attempt to capture the messages in the reports from SCOPE and IPCC as well as some other information and in a way that is accessible to the non-expert reader. The ambition is also to provide the reader with a picture of the different factors that scientists enter into their scenarios of the future. It should particularly interest policy advisors, scientists in other disciplines and teachers. The decisions called for in a global climate convention have to be made by policy makers worldwide, but the basis for those decisions is the picture painted by scientists.

M. K. Tolba
Executive Director
UNEP

J. W. B. Stewart
President
SCOPE

Acknowledgements

THIS BOOK IS A SYNTHESIS OF SEVERAL SCIENTIFIC reports, most particularly the work done by the Intergovernmental Panel on Climate Change as presented in three IPCC reports from 1990. Another important source of information has been report SCOPE 29 from the Scientific Committee on Problems of the Environment. Without the work many scientists have put into those reports, this book would not have been possible.

I also wish to thank my scientific editors, Professor Bo R. Döös and Dr. John S. Perry for providing valuable comments on the manuscript. For help in editing language and style and for support throughout the work on the book, I thank Dr. Cynthia de Wit.

Many others have contributed with information and assistance, whom I also wish to thank, in particular Professor Bert Bolin, Professor R. E. Munn and Professor Henning Rodhe.

A. Nilsson

Figures have been taken from the following sources:

Fig. 2.1 from *Saving the Ozone Layer: A Global Task*, Swedish Society for the Conservation of Nature, 1990; Figs 2.2, 3.2, 3.3, 3.12, 3.13, 4.1, 4.2, 5.1, 5.2, 5.3, 5.5, 5.6, 5.7, 6.1, 6.2, 6.3, 7.1, 7.2, 7.4, 7.6, 9.2 and 9.3 from *IPCC Scientific Assessment*, Cambridge University Press, 1990; Figs 2.3, 5.4, 7.5, 8.2 and 9.4 from *The Greenhouse Effect and its Implications for the European Community*, Commission of the European Communities, Report EUR 12707; Fig. 3.1 from *Global Ecosystems*, Macmillan, 1973; Figs 3.4 and 9.5 from *Environmental Issues Requiring International Actions*, IIASA Report; Figs 3.5, 3.6, 3.8, 3.10, 9.6, 12.1 and 12.2 from *Climate Change: The IPCC Response Strategies*; Figs 3.7, 3.9 and 3.11 from *Proceedings of the 2nd World Climate Conference*, Cambridge University Press, 1991; Figs 7.3, 8.1, 8.3 and 9.1 from *SCOPE 29*; Figs 7.7, 7.8, 11.1 and 11.2 from *IPCC Climate Change: The IPCC Impacts Assessment*, Commonwealth of Australia copyright, reproduced by permission, 1990; Figs 8.4, 8.5, 8.6 and 8.7 from *Climate Change and World Agriculture*, Earthscan Publications, London, 1990.

We are grateful to the authors and publishers of these sources for permission to reproduce them.

Chapter 1

The history of our future

Stockholm, Sweden, 30 November 2030

Dear friends,

The year is 2030 and I am leafing back though old news accounts in trying to understand the world I am living in and how it was shaped. Here in Scandinavia, we are rather well off. Industry has been able to keep up with changes by being alert at adopting energy-saving technologies. Conversion from fossil fuels to other energy sources is also paying off. There were some rough years, though, when most of the energy systems were still dependent on fuels that were hit with special carbon dioxide taxes.

There are definitely some changes compared to the end of last century when greenhouse warming was still considered an environmental controversy. The big visual change is our landscape. In the past few years, farmers' fields have taken on a new look, with fast-growing willow as one of the major crops along with a special kind of tall grass grown for fuel. Food production is of course also important, but where there used to be wheat, they now grow corn, which never used to ripen well this far north. Some farmers get bigger crops these days since the growing seasons are much longer than 30 years ago. It is only the potato farmers in southern Sweden who complain. Quite often their crops are devastated by the Colorado beetle, which seldom came this far north before. There are also some warning signs about the corn borer moving north.

One notices how climate has changed the forests, too, especially where there has been extensive logging. Even where the forestry companies have tried to plant spruce and pine in the cleared areas, the trees are not doing very well. In other places one can see areas of oak and beech getting established even as far north as Stockholm. I wonder if the big dark forest of spruce, that I remember from my childhood, will soon be replaced by deciduous trees that will paint the hillside in beautiful autumn colors.

I sit here wishing it would snow soon. I used to like skiing though the white-covered landscape, but we haven't had real winters for many, many years now. But slowly, I am learning to like it this way too. It is warmer, it is rainier, but not really much different than if I had moved south, to central Europe rather than staying in Scandinavia.

What troubles me is the global politics of today and the strife of people further south. In the end of the 1990s, news reports of flooding in the Ganges delta became more and more frequent. Now it seems like there is a new catastrophe every year, as if the sea is taking over more of the productive wetlands while the people living there have nowhere else to go. Floods sweep away houses and people during the monsoon rains and the sea is inundating farmland and fishing ground that used to sustain many families. It is the salty water that is destroying the sensitive breeding grounds for fish and shrimp. The salty water seems to contaminate everything in the coastal zone. Rice is getting harder to grow and wells with good fresh water are becoming scarce.

While summer rains have become more intensive in the south, winters seem to have dried up. We keep getting reports of poor yields from the rice paddies in India and that grain imports put a larger and larger strain on the economies of some countries in south-east Asia even if others have actually gained. And grain from the world market does not come cheap these days. Some years there just does not seem to be enough to go around and prices soar. What used to be the bread baskets of the world have been hit by several severe droughts. On the Great Plains in North America, they talk about the risk of another Dust Bowl like the one that hit in the 1930s. It is not so much the change in rainfall, although the talk among farmers is about how long they have to wait for rain, but rather the warm summer days drying out the soil. The farmers have tried to offset the thirst of the soils with irrigation, but water has become both expensive and scarce and yields keep decreasing. The corn crop last year was only three-quarters of what it used to be in the 1990s. Many Great Plains farmers are finally getting the message and starting to grow drought-tolerant wheat instead, which does better than corn.

One would think Canada, being further north, would have benefited from the warmer climate, but even there lack of soil moisture is taking its toll on the yield of grain. In countries around the Mediterranean, the dryness is taking a hard toll on the soil and some areas are no longer usuable for agriculture. One country that has been able to take advantage of the warmer weather is Russia, but in general it seems that it has taken too long to adjust agricultural practices to the new climate.

We probably could have done better on a global scale, if we had been willing to recognize what was coming. Granted, no one really knew how much it would rain and where, how hot summer days would become, how cold or mild the winter. But we had ample warning that there was something in store. I remember some of the catastrophic scenarios painted in the popular press during the 1990s, of whole island nations disappearing under the rising sea level because of the expanding oceans and melting ice sheets. It was difficult to know what to believe. How sure was the scientific community? Could their models of global climate change be trusted? They were, after all, only models, and no one could tell if our mild winters in Scandinavia or the dry summers of the Great Plains were part of a normal variation or if something was actually happening.

Yes, we finally acted, slowly cutting down on emissions of greenhouse gases. But it took a long time before we really addressed the global politics of the changes ahead. The result I see now, looking back, is a few winners, Scandinavia being one with increased agricultural yields and more productive forestry. But I also see many losers, most of them in the south where food or productive land was scarce to start with. It is still not too late to change things, to address the global inequities that have become accentuated because of changes in climate. But it would have been easier if we had started earlier, even if we did not have all the answers. We still do not have them, we can only guess what is in store for the next 40 years when we try to shape our future.

With warm regards,

Annika Nilsson

The year is now 1992 and we still do not have the answers to all our questions on what will happen to global and regional climate in the future. The picture might be much brighter than the one painted above, but it might be gloomier. Nevertheless, we have some answers from the information that scientists from different disciplines have gathered so far and from the models which they use to analyze the future. And their work started more than a century ago.

AN EMERGING SCIENCE

Atmospheric science as it regards climate warming can be traced back to the 1820s when the Frenchman Fourier first described the greenhouse effect: ''The atmosphere acts like a hot house, because it lets through the light rays of the sun but retains the dark rays from the ground.'' In 1861, the greenhouse gases were mentioned for the first time when Tyndall, in England, measured the absorption of heat radiation by water vapor and carbon dioxide. There was no implication that the strength of the greenhouse effect would change, however. That came later when the Swede Arrhenius wrote a paper in 1896. He tried to correlate changes in surface temperature of the Earth with changes in atmospheric carbon dioxide to explain the occurrence of ice ages. Seven years later he noted that industry might put out enough carbon dioxide to actually warm the Earth.

Arrhenius was followed by the Englishman Callendar who in 1938 published a paper on the artificial production of carbon dioxide and its influence on climate. Callendar tried to explain the temperature increases that were occurring at the time, but in the 1940s the temperature trend changed, the northern hemisphere started to cool and his work was no longer immediately relevant.

The early discussion of the greenhouse effect was of concern only for the scientific community. In 1955, however, the Hungarian-American von Neuman in a popular article posed the question: "Can we survive technology?". One of his points was climate change. He was followed by the American scientist Plass who in 1959 wrote the first popular account of the carbon dioxide theory of climate change.

Meanwhile, science continued to build a groundwork of understanding, and by the mid-1960s the discussion had reached the political level. The Conservation Foundation held a conference airing concerns and a US government report about a wide range of environmental problems identified most of the issues in today's discussion of carbon dioxide and climate warming.

In the 1970s, the tone changed and global cooling became the issue of concern with the focus on aerosol emissions from industry. But the idea that people could influence the global climate was taking hold. It was accentuated by the first reports on CFCs and their potential to destroy the layer of stratospheric ozone that protects Earth from ultraviolet radiation. And more was in store. In 1974, Manabe presented the first computer models of climate and climate change, which alerted many scientists to the fact that human activities could also affect climate on a global scale. At the First World Climate Conference in Geneva in 1979, the issue of climate warming once again came into focus.

LAYING THE FOUNDATION

There were many uncertainties about how the increase in carbon dioxide would actually affect the climate and there was an acute need to critically evaluate the scientific base for any doomsday prophesies. The United Nations Environment Programme (UNEP), the World Meteorological Organization (WMO) and the International Council of Scientific Unions (ICSU) thus held an expert meeting in Villach, Austria in 1980. The conclusion was that carbon dioxide-induced climate change indeed was a major environmental issue, but that it would be premature to develop a management plan to control carbon dioxide levels. More research was needed to create a firm scientific base. Already at this meeting, the need for cooperation between developing and industrialized nations was taken up.

This scientific review was followed by several others, with similar conclusions. Meanwhile, the work on atmospheric models continued to provide new information. The US Environmental Protection Agency (EPA) already in 1983 concluded that only a ban on coal-use, instituted before 2000, would effectively slow down the rate of global temperature change and delay a 2°C increase until 2055. They also said that such a ban would probably be politically and economically infeasible.

In 1985, the international scientific community had once

again gathered at Villaeh in Austria. This time they had a firmer scientific base on which to base their conclusions. The Scientific Committee on Problems of the Environment (SCOPE) had commissioned a review, resulting in the report "The Greenhouse Effect, Climate Change and Ecosystems", which clearly established increases in greenhouse gases as an international problem.

Further scientific meetings in 1987, along with accompanying press coverage gave the issue world-wide attention. The discovery of the ozone "hole" over Antarctica and international agreements to limit emissions of CFCs further accentuated the need to deal with the global politics of the atmosphere.

During this time, work on climate models had improved considerably, which provided a base for policy discussion to complement the description of the problem. At the 1988 Toronto Conference on the Changing Atmosphere, it was thus time to start formulating political goals. The discussion of energy, food security, industry, forecasting, legal dimensions etc, resulted in a call for a reduction of carbon dioxide emissions by approximately 20 percent of 1988 levels by the year 2005, as an initial global goal. It did not become anything more than a call, however.

Much of the international work since then has been co-ordinated by the Intergovernmental Panel on Climate Change (IPCC) under the auspices of the World Meteorological Organization and United Nations Environment Programme. In 1990, this panel presented a consensus report based on assessments of science, impact and policy made by more than a hundred scientists. The previous conclusions were confirmed — global climate change is a real problem which has to be dealt with. The next step is to reach an international agreement on how to go ahead and to sign a world climate convention at the international conference on environment and development in Rio de Janeiro, Brazil in 1992.

Chapter 2

The greenhouse Earth

A FUTURE OF CLIMATE WARMING IS OFTEN described as an increase in the greenhouse effect. It is easy to envision a glassed-in building trapping the heat of the sun, creating a climate that is quite different from the surroundings. The Earth with its atmosphere acts very much like a greenhouse. Without our atmosphere, the climate on the surface of the Earth would be very harsh with nothing to prevent the planet's heat from escaping into space. Instead of glass, Earth is surrounded by gases, and among them are gases that have an ability to trap heat. (Figure 2.1)

One of the major greenhouse gases is water vapor. Its effect can be studied when comparing the climate in an environment with moist air with that of a dry desert, for example, the coastal zones of south-east Asia with the climate in the Sahara desert. In south-east Asia where the water-vapor content of the atmosphere is high, the temperature stays reasonably constant regardless of whether it is day or night. In the desert, on the other hand, the temperature can drop quite drastically once the sun has set.

Other greenhouse gases are carbon dioxide, methane, nitrous oxide and chlorofluorocarbons (CFCs). It is these gases that are the cause for concern when talking about man-made climate warming and climate change. The concentrations of all of them are increasing in the atmosphere and the prediction is that their accumulation is enhancing the natural greenhouse effect. It will become a warmer greenhouse because the gases will be even more effective in trapping the heat trying to leave the atmosphere.

A planet is, however, both larger and more complex than an ordinary greenhouse and the atmosphere does not evenly spread an increase in temperature. Rather, the Earth with its atmosphere can be described as a playground for weather systems, where air, water and land interplay to determine the climatic patterns in different regions. If the North

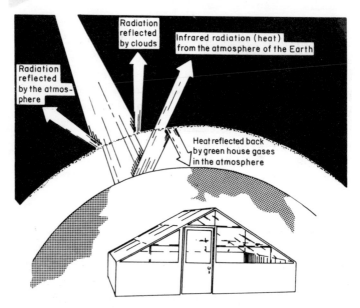

Fig 2.1.   The greenhouse earth

Atlantic low pressure path moves a little north or south, this might not seem like a big change on the global scale, but for the weather of Northern Europe it is what determines the number of clear and rainy days or if there will be real winter weather. A shift in the monsoon further north might bring more rain to the dry sub-Saharan region, but it is difficult to predict how much and when. And no one knows how fast the rain water will evaporate as the temperature may also increase. This complexity is one of the reasons scientists are still reluctant to make regional predictions about the effects of climate warming, even when they are quite confident about the global trends.

My imagined reflection looking back from the year 2030 is one possible future based on some of the regional studies that have been made in spite of the uncertainties. It is based on what is usually called a business-as-usual scenario, in which emission of greenhouse gases will continue to grow at the present rate. In that scenario, the amount of carbon dioxide in the atmosphere will have doubled in 2025 resulting in a global temperature increase of somewhere between 1.5 and 4.5°C, but most likely in the lower end of this range of uncertainty. Regardless of the exact numbers, the change in temperature is of the same magnitude as the change in global temperature from the last ice-age to today.

TODAY'S CLIMATE

Drought and famine have been hitting sub-Saharan Africa since the 1970s. The pictures of starving people trying to find more fruitful land to grow their crops and feed their animals are becoming all too familiar. A nagging question is whether it is only population growth and war making the situation worse or whether the landscape is becoming drier, less fruitful. Is the desert expanding because of a change in climate?

In 1991, television and newspapers brought pictures of catastrophic floods in Bangladesh. In the speculations about the causes, one could read about deforestation on the slopes of the Himalayas—no trees to catch and ease the flow of water. But the rains were also unusually heavy. Was that only the result of an unusually severe monsoon, or was it an indication of something fundamentally new?

In the United States, several hot and dry summers in a row in the Midwest have brought speculation about this being the first manifestations of a warmer world. In Scandinavia, lack of snow during the past winters is hitting the tourist industry hard. Will the cold winters return or is this a sign of new times?

As one looks at climate in the past years, it is easy to find unusual weather, but none of the examples above can with any certainty be attributed to a trend of climate warming. This is in spite of the fact that the past 10 years brought the seven warmest years on record and that the mean global temperature has increased between 0.3 and 0.6°C during the past 100 years. The problem in detecting climate change is that the unusual is easily masked by natural year-to-year variation in the weather.

Neither is establishing trends in global temperature as simple as reading the window thermometer. To be of value, one needs long time trends and temperature measurements that can be validated. Detailed land-based measurements started in the 1850s and since the beginning of this century, there have also been marine data. In spite of difficulties in excluding faulty or misleading temperature data, it is possible to see some patterns and there are independent data sets that show the same trends: one warming period between 1890 and 1940 and a second since 1970 (Figure 2.2). The temperature has not increased evenly over the globe and some regions in the northern hemisphere have even cooled until recently. This underscores a point often made, that a future increase in global temperature will not affect all regions equally.

In addition to direct measures of temperature, there is the continued retreat of valley glaciers that can be observed globally. In the Swedish National Park of Sarek one can follow the yearly painted marks on rocks below some glaciers that have retreated about a kilometer during the past century. Part of this could be a continuation of retreat in response to the warming after The Little Ice Age (approximately 1400–1800), but anthropogenic factors may also be involved, looking at the global trend. Another sign of warmer climate is the patterns in the temperatures one gets from boreholes in permafrost areas. In recent years

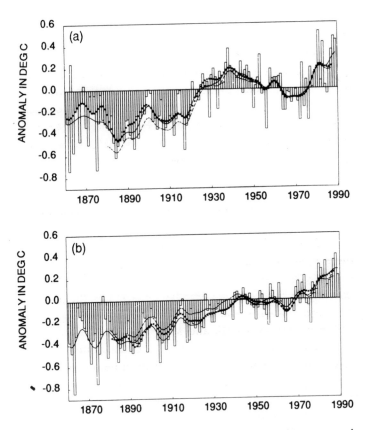

Fig 2.2. Land air temperatures from the mid-1800s, expressed as anomalies to 1951–1980. (a) Northern Hemisphere, (b) Southern Hemisphere. 1990 and 1991 brought another two unusually warm years (not in the graph).

it has also been possible to record a real decrease in snow cover in the Northern Hemisphere. In the tropics there are signs that higher water temperatures have led to an increased bleaching of coral reefs.

Temperature statistics since the mid 1800s together with these signs tell the story of increasing global temperature.

What is not possible to say is whether this increase is man-made or part of a natural variation in climate. Scientists can only correlate it with increases in greenhouse gases, not prove any cause and effect relationship. There are some scientists who point at the importance of looking at longer time series than those available from direct temperature data. Those changes can be studied by using ice core data that go back as much as 160,000 years. In addition, there are variations in tree rings, which record changes in summer temperature during the past millennium. These fluctuations are larger than the changes during the past century. Unfortunately, they cannot be directly compared with the temperature record from the past century, as the resolution is not good enough (Figure 2.3).

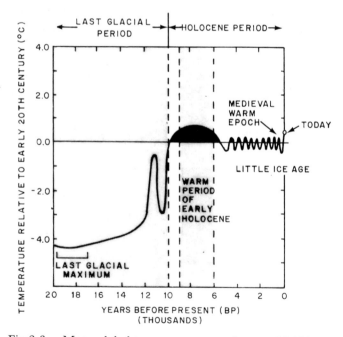

Fig 2.3.   Mean global temperature over the past 20,000 years

It is, however, not primarily temperature trends on which scientists base their concern for climate warming. It is rather the understanding of physical laws governing how incoming solar radiation is trapped as heat by greenhouse gases that has made them sound the alarm. Knowing the concentration of carbon dioxide, methane, nitrous oxide and chlorofluorocarbons in the atmosphere, one can calculate how effective a heat trap the Earth's atmosphere is. Knowing the trends of increases in concentration of these gases makes it possible to estimate increases in global temperature. And the estimates of temperature that are made in the theoretical models of an increased greenhouse effect are consistent with the trends that can be studied by direct measurements.

There have also been attempts at finding changes in precipitation and there are some regional patterns, such as a decrease in precipitation in Africa and increases in Europe, western Asia and India. There is no evidence, however, of any global trends. The great variation in precipitation in combination with the often poor quality of the data also make it difficult to be sure about trends and scientists have not been able to correlate any changes with an increased greenhouse effect.

Chapter 3

The greenhouse gases

THE GREENHOUSE GASES HAVE ONE THING IN common: they can trap heat. In other respects, they are very different. They have different sources and different sinks. Some are part of a natural circulation where human intervention can be small or large. Others are purely man-made. Before comparing them, it might be useful to get to know them one by one.

CARBON DIOXIDE (CO_2)

As you breathe, a few more molecules of carbon dioxide are added to the atmosphere. The same is true for cattle, fish, wild animals, bacteria or any breathing creature. This gas is eventually reabsorbed by trees, bushes and phytoplankton. Using photosynthesis, they catch it and store the carbon as a source of energy. Looking at life globally, a large part of Earth's carbon is stored in biomass and only a portion of it is converted to carbon dioxide gas when people and animals eat and metabolize the biomass. Some of the stored carbon is returned to the atmosphere by natural decomposition by bacteria and other organisms in the soil. In general, there is a balance between the ability of vegetation and plankton to catch carbon dioxide and the release of the gas from metabolism and decomposition. (Figure 3.1)

Trees, bushes and other sources of biomass have also been used for burning in a more literal sense, to provide heat for cooking and comfort. As long as new trees and bushes grow at the same rate as old ones are cut down for fuel, there is still a balance. The amount of carbon dioxide in the atmosphere remains constant.

The real problem is the use of non-renewable sources of carbon: the fossil fuels—oil, coal and gas. These fuels do not regenerate and when burned they only add to the

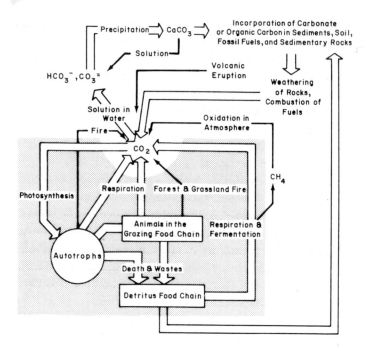

Fig 3.1. The carbon cycle. Organic phase shaded

carbon dioxide content of the atmosphere. Since the industrial revolution, we have thus added carbon dioxide to the atmosphere at an increasing rate. This is the major source of the carbon dioxide problem. In addition, a major sink of carbon, the forests, are destroyed, further offsetting the balance.

THE TRACKS OF IMBALANCE

Before the industrial revolution, the carbon dioxide concentration in the atmosphere was somewhere between 265 and 280 molecules of carbon dioxide per million other molecules in the air (265–280 ppm). Since then it has increased to about 353 ppm. The trends can be followed by measuring the carbon dioxide-content in air bubbles trapped in ice cores. Measurements from Siple Station, Antarctica show that the carbon dioxide-concentration started to rise around 1800 and had increased about 15 ppm by 1900. Since 1958, there are also direct measurements in air made at Mauna Loa, Hawaii. The rate of increase is now 1.8 ppm per year. (Figures 3.2 and 3.3)

The increase in carbon dioxide we see today is not a direct measure of the imbalance between biomass accumulation and biomass use. The picture is complicated by the oceans, which can dissolve the gas and convert it to other forms of carbon. The oceans are in fact Earth's largest reservoirs of carbon. Their ability to dissolve the gas provides for a time lag on any imbalance in the release and use of carbon dioxide by living organisms.

The reservoir capacity of the oceans is a function of several different processes and is not completely understood. The water temperature and concentration of the gas in the surface water compared to the air is one factor that determines whether more gas will be dissolved, but equally important are the currents that provide for mixing of surface waters with the deep ocean. In addition, there is a biological pump in the form of carbon dioxide-catching plankton that move carbon to the deep sea floor as a rain of detritus.

What is the net result of all the additions and exchanges of carbon to the atmosphere? According to one estimate, burning of fossil fuels, deforestation and changes in land use from 1850 to 1986 add up to an input of 312 billion tons of carbon (312 GtC). This corresponds to about 41 percent of the carbon that has been added to the atmosphere. The rest has been stored, at least temporarily, by sinks such as the ocean.

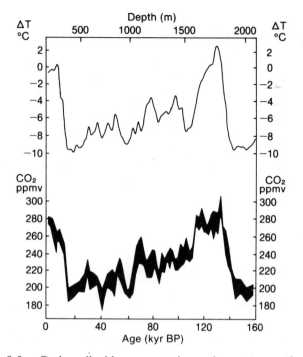

Fig 3.2. Carbon dioxide concentration and temperature change during the past 160,000 years, as determined from the ice core from Vostock, Antarctica

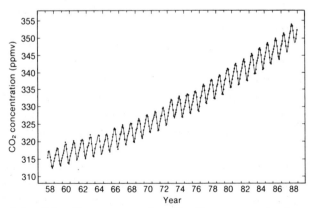

Fig 3.3. Monthly average carbon dioxide concentration in parts per million of dry air measured at Mauna Loa, Hawaii

A question is whether the oceans are the only sinks. Based on current knowledge of their capacity to store carbon dioxide, the calculations do not come out balanced. There is more carbon emitted than can be found in the atmosphere and oceans together according to the models of the oceans used today. But all attempts at identifying a missing sink in the oceans have so far failed. A terrestrial sink might be an increased capacity of vegetation to store carbon due to fertilization. There are also suggestions that the North Atlantic acts as a more efficient sink than previously thought.

EMISSION OF CARBON DIOXIDE

The use of fossil fuels started in the 1800s and today we see the effects as increased concentrations of carbon dioxide in the atmosphere—the carbon dioxide problem. At present, about 6 Gt carbon is added each year from fossil fuels and other industrial sources such as cement manufacturing. This accounts for a little less than two thirds of all carbon dioxide that has been added to the atmosphere since the mid 1800s.

There is no question that the increase will continue. In spite of both industry and transportation becoming more energy efficient, we burn more fossil fuel every year. Individual consumption increases steadily, so far mostly in the industrialized world but increasingly also in some developing countries. Also, world population continues to grow.

General statements about future trends are easy to make. It is considerably more difficult to quantify future emission of carbon dioxide as there are no constant correlations between economic growth, population and energy use. Rather, the relationships depend to a large extent on energy

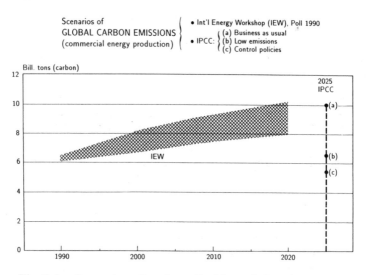

Fig 3.4. Scenarios of carbon dioxide emission from energy production vary greatly depending on what policy options are entered into the calculations

prices, policy, technological developments in energy efficiency and choice of energy source.

There are a number of scenarios of future carbon dioxide emission (Figures 3.4 and 3.5). The result is a range of possibilities looking at the future. The upper bound shows an increase of emission to more than 20 billion tons of carbon a year compared to a 1986 level of about 5 billion tons. The lower bound is roughly the present level, or even lower.

This range of estimates in the scientific data does not mean that nothing can be said about the future. Rather, it is a sign of the options available to decision makers. A more detailed analysis of the studies shows, for example, that global energy productivity and efficiency are currently far from optimal and can be improved considerably.

Several studies also show that strategies to limit the increase of greenhouse gases to a doubling would still allow

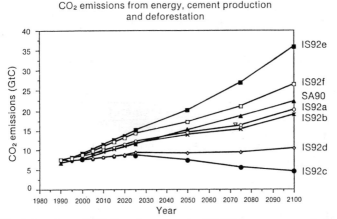

CO₂ emissions from energy, cement production and deforestation

Fig 3.5. In a 1992 Supplement, the IPCC has presented six different scenarios of carbon dioxide emissions instead of the one business-as-usual scenario on the original report (called SA90 in the figure). The major differences compared to previous estimates is that higher population forecasts increase the emission estimates, while phase-out of halocarbons and more optimistic renewable energy costs reduce them. The small difference between a and b is mainly accounted for by the commitment to stabilize or reduce carbon dioxide emission made by many OECD countries

for economic growth in developing countries, but it requires political decisions about our energy future. We have to make fundamental changes in lifestyle and technology, especially in the industrialized nations. If we continue with business as usual, the carbon dioxide in the atmosphere will increase considerably. How much carbon dioxide that will enter the atmosphere before there is a balance in the carbon cycle depends to a large extent on how rapidly and how stringently we are willing to cut emissions.

DEFORESTATION   *Sự já rừng .*

Emission of carbon dioxide is only one part of the carbon cycle. At the other end is the vegetation that accounts for the assimilation of carbon dioxide from the atmosphere and the binding of carbon in biomass. This end of the equation is no less problematic than the emission side as deforestation and land-use changes have diminished the global storage of carbon as well as the capacity to bind carbon dioxide.

What share of the increase in atmospheric carbon dioxide can be attributed to deforestation and land-use changes? The basis for the question is the fact that vegetation and soils from unmanaged forests hold 20 to 100 times more carbon per unit area than agricultural land. As forests are cut back to clear land, more carbon is released into the atmosphere. In the past, most of this land clearing was done in the temperate zone. Now the major source of carbon is deforestation in the tropics, which has increased significantly since the 1950s.

Between 1860 and 1980, deforestation and land-use change have added somewhere between 80 and 150 billion tons of carbon (GtC) to the atmosphere. This is a little more than one third of the carbon dioxide that has been added to the atmosphere in the past century. The release from tropical forests is estimated to be 2–3 times greater than that from middle and high latitudes. The yearly global input is somewhere between 0.6 and 2.6 GtC.

The difficulties in making these estimates are many. There is a lack of basic information both on the amount of forest land, today and in the past, and the carbon dynamics of different vegetation systems. For example, estimates of carbon storage in the tropical forest depend on how dense the forest is and many areas that have previously been considered to be covered by closed forests should rather be called semi-closed. Secondary forest in the tropics, i.e. new growth after a clearing, stores less carbon than natural tropical forest, which has to be taken into account when estimating the total storage capacity.

Carbon storage in the soil is also important and

dependent on the vegetation type. When natural ecosystems are converted to agricultural land, a large proportion of the soil carbon can be lost as plants and dead organic matter are removed. One way of estimating total carbon storage is to use vegetation maps and reassessment of such map studies also point to some of the difficulties in estimating carbon flux. For example, forests in the former Soviet Union used to be considered net carbon sinks, but it now seems clear that the opposite is true. Forests at mid-latitudes elsewhere in Europe and North America most likely store more carbon now than a century ago.

The calculations of the contribution of forests to the increasing greenhouse effect also have not taken into account any effects of air pollution. This may be important in decreasing carbon storage, thus adding carbon dioxide, but the magnitude of any effect is poorly known. There are also examples of poorly quantified processes that can remove carbon from circulation. One is charcoal formation during biomass burning. Another is that carbon dioxide has a fertilizing effect on vegetation, which might increase its capacity to store carbon.

The relative role of deforestation in the tropics and the emissions of carbon dioxide from the burning of fossil fuel are sensitive issues in international politics of the atmosphere as it influences decisions on where changes have to be made. Just one example is the Indian environmental group Center for Science and Environment, which has called into question estimates of deforestation rates made by the World Resources Institute. Regardless of who is right, it is safe to say that there is much uncertainty in what exact numbers to use in the different estimates of carbon dioxide emissions.

In spite of all the uncertainties, it is clear that deforestation and the use of fossil fuels add more and more carbon dioxide to the atmosphere. To stabilize the atmospheric concentration of carbon dioxide at the current level requires a reduction of global emissions by at least 60 percent.

METHANE (CH_4)

The swamp is filled with water. It does not look very habitable, yet it is teeming with life. Even the areas void of oxygen are alive. They are the home of anaerobic bacteria. A major activity of the anaerobic bacteria is to extract energy from dead organic matter. They are not as efficient as their oxygen-breathing cousins, who convert the organic matter to carbon dioxide and water. Instead, their metabolic process stops a little short, leaving the gas methane as a major product.

The same bacterial process can be found in bogs in the polar regions and in rice paddies. It is also an important part of the digestion of food by grazing animals and the methane-producing bacteria can be found in the gut of these animals as well as in the guts of termites. Landfills also contribute to the increasing methane concentration in the atmosphere, by setting up conditions that favor methane-producing bacteria (Figure 3.6).

Once methane is released into the atmosphere, it enters a complex series of chemical reactions that eventually break it down to carbon dioxide and water. Before it is broken down, however, it may stay in the atmosphere for eight to eleven years and there is thus a store of trace amounts of methane in the air.

Methane shares an important property with carbon dioxide. It transmits light but blocks the transmission of heat trying to leave the atmosphere. It is thus a greenhouse gas. Since the start of the current detailed measurements of methane in the atmosphere in 1975, the concentration

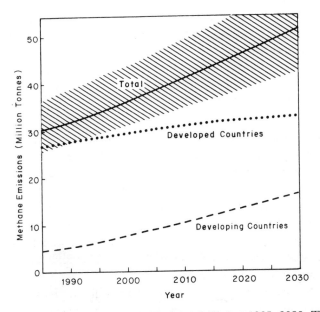

Fig 3.6. Methane emissions from land-fill sites 1985–2030. The cross hatched area indicates range of uncertainty

has increased at a rate of more than 1 percent per year. Earlier data from air bubbles in the ice sheets of Greenland and Antarctica show an exponential increase of methane in the atmosphere during the past 300 years. The increase correlates well with the increase in human world population, which gives a clue that the increase is caused by human activities, primarily agriculture (Figure 3.7).

Another source of methane in the atmosphere is connected to the use of fossil fuels. The gas is released in the exploitation of methane as natural gas, from leaking pipelines and from coal mining. Incomplete combustion of fossil fuels or biomass also adds methane to the atmosphere.

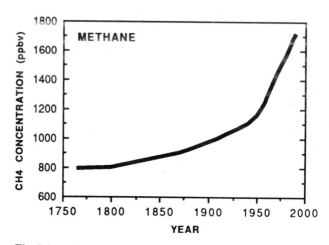

Fig 3.7. Concentration of methane in the atmosphere

It is very difficult to make good estimates of emissions from the different sources. The only source that is relatively well quantified is the release from large herbivores, such as cows, where one can take into account the metabolic system of various species of domestic and non-domestic animals and the number of animals world-wide. Some of the numbers have recently been challenged by scientists who claim that the skinny cows of many developing countries cannot be compared to the well-fed animals in for example the United States. Small biomass eaters, such as termites, may be almost as important as large herbivores in adding methane to the atmosphere, even if the numbers are much more uncertain.

Quantifying emissions from rice paddies has proven difficult. The emissions vary with amount of land in cultivation and also depend on fertilization, water management, density of rice plants and other agricultural practices. Measurements from laboratory experiments and from actual rice fields in Europe differ considerably.

Table 3.1   Estimated sources and sinks of methane

	Annual release (Tg CH$_4$)	Range Tg CH$_4$
Source		
Natural wetlands (bogs, swamps, tundra, etc.)	115	100–200
Rice paddies	110	25–170
Enteric fermentation (animals)	80	65–100
Gas drilling, venting, transmission	45	25– 50
Biomass burning	40	20– 80
Termites	40	10–100
Landfills	40	20– 70
Coal mining	35	19– 50
Oceans	10	5– 20
Freshwaters	5	1– 25
CH$_4$ hydrate destabilization	5	0–100
Sink		
Removal by soils	30	15– 45
Reaction with OH in the atmosphere	500	400–600
Atmospheric increase	44	40– 48

IPCC Scientific Assessment, Cambridge University Press, 1990

New measurements from East Asian rice paddies vary greatly, but generally indicate that emissions from rice paddies should be revised downward compared to the numbers in the 1990 IPCC report.

Emission figures from swamps and marshes have previously been speculative, as most of them have been based only on mid-latitude wetlands, which may differ very much from a tropical swamp. However, in the past few years new data have given a much better base. The data support earlier numbers but reverse the relative importance of tropical and high latitude systems.

Table 3.1 adds up the emissions and the ranges of uncertainty, which show how difficult it is to put exact numbers on emission. Figure 3.8 shows how the relative emissions are projected to change.

Not all of methane stays in the atmosphere. Processes such as atmospheric oxidation of the methane into carbon dioxide and water and decomposition in the soil remove some of it. The rate at which methane is removed from the atmosphere depends on other pollutants. For example, increases in emission of carbon monoxide, which is not a greenhouse gas itself, can contribute to increases in methane and to climate warming.

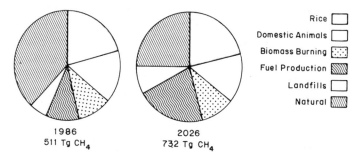

Fig 3.8. Emissions of methane by source in 1986 and as projected by 2026

A LEVELLING INCREASE

Even with the difficulties in quantifying the methane emission from various sources, there is no doubt that methane emissions have increased over the past 40 years. Between 1960 and 1975, cattle production, harvested rice paddy area and the consumption of natural gas have increased exponentially. These increases are now levelling off, but the current atmospheric concentration of methane is greater than any time in the past 160,000 years.

What will happen to methane emissions in the future? The answer depends on several social, economical and political factors. Population growth will mean more land used for rice cultivation and an increase in the number of methane-producing farm animals. New varieties of rice plants, more efficient use of mineral fertilizer and other adjustments in agricultural practices could act in the opposite direction. A third unknown is the capacity of atmospheric sinks for methane, which will depend on future emission of other air pollutants.

Methane emission will also be influenced by the future climate. If soil moisture increases, wetlands will produce more methane and if there is a decrease in precipitation and soil moisture, methane flux will decrease. Most sensitive to climate change are the tundra regions, where warmer temperatures will accelerate the biological activity in the soil. If large areas of permafrost thaw, the methane flux could increase drastically.

A reduction of global man-made emissions by 15 to 20 percent is needed to stabilize the atmospheric methane concentration at the present level.

NITROUS OXIDE (N_2O)

Many farmers spread nitrogen fertilizers in their fields to enhance the growth of their crops. Most of the nitrogen is taken up by the crop, but some is leached into surface water and ground water. Some of it enters the atmosphere. As with methane, the flux of nitrogen depends on the microbial activity in the soil. If the fertilizer is applied as ammonia, for example, bacteria convert some of it to the gas nitrous oxide. Nitrogen in this particular form serves as a greenhouse gas when it has entered the atmosphere.

How much of the nitrogen is lost to the atmosphere directly or indirectly via bacterial production of nitrous oxide in water? To a large extent, it depends on agricultural practices, such as plowing and irrigation, and on temperature, soil type and weather. Another effect of agriculture on the concentration of nitrous oxide in the atmosphere is in the breaking of new land when nitrogen bound in the soil and vegetation is released. As the world population has increased, so has the area of cultivated land and the nitrous oxide emission from this source has doubled between 1930 and 1980 (Figure 3.9).

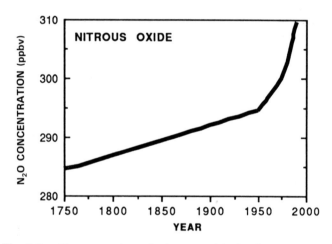

Fig 3.9. Concentration of nitrous oxide in the atmosphere

Biomass burning and burning of fossil fuels have previously been considered significant sources of nitrous oxide to the atmosphere, but a re-evaluation of the procedure used for the measurements has changed that view. Table 3.2 gives the natural and anthropogenic sources

Table 3.2   Estimated sources and sinks of nitrous oxide

	Range (Tg N per yr)
Source	
Oceans	1.4 – 2.6
Soils — tropical forests	2.2 – 3.7
— temperate forests	0.7 – 1.5
Combustion	0.1 – 0.3
Biomass burning	0.02– 0.2
Fertilizer (including ground-water)	0.01– 2.2
TOTAL	4.4 –10.5
Sink	
Removal by soils	?
Photolysis in the stratosphere	7–13
Atmospheric increase	3–4.5

IPCC Scientific Assessment, Cambridge University Press

of nitrous oxide that contribute to the increasing concentration in the atmosphere and figure 3.10 show their relative contribution now and in the future. In the past year, three more sources have been added to the list: nylon production, nitric acid production and cars with three-way catalysts.

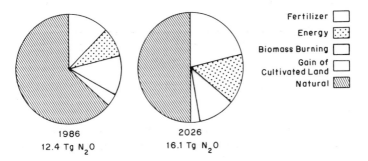

Fig 3.10.   Nitrous oxide emissions by source in 1986 and as projected by 2026

Nitrous oxide is fairly stable in the atmosphere and can stay there for about 150 to 170 years. The gas is destroyed when it is exposed to light and excited oxygen atoms in the stratosphere. In these reactions it takes part in breaking down the ozone layer, which protects the Earth from damaging ultraviolet radiation. A reduction of nitrous oxide emissions is therefore necessary both to reduce climate warming and to protect the ozone layer. An immediate reduction by 70 to 80 percent of the additional flux of nitrous oxide that has occurred since the pre-industrial era is necessary.

CFCs AND OTHER HALOCARBONS

Chlorofluorocarbons (CFCs) differ from other greenhouse gases in that they are exclusively man-made. From the 1930s, they have been produced by industry and used as a cooling medium in refrigerators and air conditioners. They are also used as a propellant gas in spray cans, as foaming agents in the production of insulation and padding and as a solvent for cleaning. The related bromofluoro-carbons are used in fire extinguishers (Table 3.3).

The halocarbons were brought to the attention of atmospheric scientists in the mid 1970s, when one started to understand that CFCs were extremely stable and that they could be transported to the stratosphere. Once there, ultraviolet light can break loose chlorine and bromine atoms, which in turn attack the ozone layer. As more and more scientific evidence as to their destructive potential has come in, including the appearance of the ozone ''hole'' over Antarctica, politicians world-wide have come to recognize the need to ban these chemicals. Meanwhile, their potential as greenhouse gases has also been recognized and has provided yet another incentive to get rid of them.

Table 3.3   Halocarbons controlled by the Montreal Protocol

Product	Chemical formula	Major use	Ozone Depletion Potential*	Greenhouse Warming Potential*
CFC-11	CCl_3F	Foaming agent Refrigeration Cleaning Aerosols	1.0	1.0
CFC-12	CCl_2F_2	Refrigeration Foaming agent Aerosols	0.9–1.0	2.8–3.4
CFC-113	CCl_2FCClF_2	Cleaning Refrigeration Foaming agent	0.8–0.9	1.3–1.4
CFC-114	$CFClF_2CClF_2$	Refrigeration Foaming agent Aerosols	0.6–0.8	3.7–4.1
CFC-115	$CClF_2CF_3$	Refrigeration	0.3–0.5	7.4–7.6
Halon-1301	CF_3Br	Fire fighting	7.8–13.2	
Halon-1211	CF_2ClBr	Fire fighting	2.2–3.0	
Halon 2402		Fire fighting	5.0–6.2	

Other ozone-depleting substances

Product	Chemical formula	Major use	Ozone Depletion Potential*	Greenhouse Warming Potential*
Carbon tetrachloride	CCl_4	Cleaning	1.0–1.2	0.34–0.35
Methyl chloroform	CH_3CCl_3	Cleaning	0.10–0.16	0.022–0.0026

Some chemical alternatives to CFCs

Product	Chemical formula	Major use	Ozone Depletion Potential*	Greenhouse Warming Potential*
HCFC-22	$CHClF_2$	Refrigerant	0.04–0.06	0.32–0.37
HCFC-123	$CHCl_2CF_2$	Refrigerant Foaming agent Cleaning	0.013–0.022	0.017–0.020
HFC-134a	CH_2FCF_3	Refrigeration	0	0.24–0.29

*According to UNEP Synthesis Report (November 1989)

Over the past few decades, the concentration of halocarbons in the atmosphere has increased more than any other greenhouse gas (Figure 3.11). To stabilize and eventually reduce the concentration of these compounds

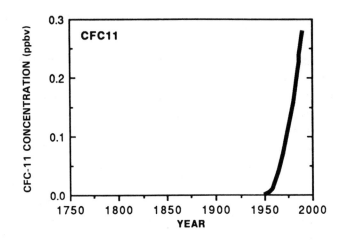

Fig 3.11. Concentration of CFC-11 in the atmosphere. This compound and the other man-made CFCs were not present at all in the atmosphere before the 1930s

will require a 70 to 95 percent reduction depending on what compound one talks about. This will probably be achieved through the international agreements that have been reached in the Montreal Protocol on Substances that Deplete the Ozone Layer. The revised version from 1990 calls for a complete phase-out of the most destructive CFCs by the year 2000 and many countries now discuss even earlier phase-outs.

The regulated CFCs will to some extent be replaced by HCFCs, which are compounds in which a hydrogen atom has been added to the molecule to make it less stable. The HCFCs are not as destructive to the ozone layer but are still greenhouse gases and despite the Montreal Protocol, halocarbons will continue to be important for the greenhouse effect for quite some time.

STRATOSPHERIC OZONE

Stratospheric ozone received attention with the discovery of the re-occurring ozone "hole" over Antarctica every spring, which first alerted politicians internationally that we are able to affect the environment on a global scale. Ozone, however, has many roles in the atmosphere and it also enters into climate warming. It is a greenhouse gas itself, as well as an important actor in the chemistry of the atmosphere. Furthermore, its effect sometimes depends on its height in the atmosphere, as it can block both solar radiation and heat from the Earth.

The decrease in stratospheric ozone that has been recorded globally leads to an increase in incoming radiation and thus a warming of the troposphere. A competing trend is that the decrease in ozone leads to less absorption of incoming sunlight. This cools the stratosphere and will also lead to a cooling of the troposphere and the surface of the Earth as the stratosphere has less heat to share. The balance of the two processes depends on atmospheric chemistry involving future emission of a number of compounds, but scientists lack the data to predict how tropospheric temperatures can be affected.

Recently, new information focusing on the lower stratosphere has been presented. Here, the decreasing concentrations of ozone have alerted scientists about the importance of stratospheric ozone as a greenhouse gas in itself. The decreases in ozone that have been recorded apart from the ozone "hole" over Antarctica might be large enough in the Northern Hemisphere to counterbalance the warming caused by increases in halocarbons. This could be part of the explanation why global temperature changes have not been as large as predicted by climate models.

44

TROPOSPHERIC OZONE

Ozone is also present closer to the Earth's surface, in the troposphere, where it is very unevenly distributed. It is produced in photochemical reactions involving gases such as methane, other hydrocarbons, carbon monoxide and nitrogen oxides, all of which are formed in combustion processes and emitted from cars, for example. The higher concentrations are found over industrialized areas with extremes where photochemical smog is formed. In Europe, ozone values were about twice as high in the 1970s as between 1930 and 1950 (Figure 3.12).

Biomass burning also creates conditions that are favorable for ozone formation in the troposphere. There are no long-term records of tropospheric ozone concentration in the

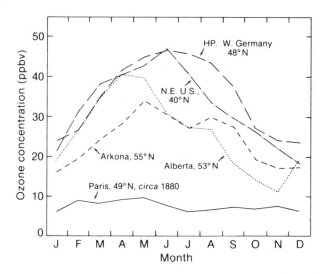

Fig 3.12. The seasonal variation of tropospheric ozone. The data are from the 1970s and 80s, which can be compared to the measurement from Paris in the late 1800s

Southern Hemisphere, but there seems to be an increase in areas with extensive biomass burning such as South America, Africa and India.

As with stratospheric ozone, it is very difficult to model future changes in ozone concentration in the troposphere. The dynamics of the chemical reactions involved are not well understood and they may also be influenced by climate, such as changes in cloud cover, precipitation and circulation patterns.

AEROSOLS

On a hazy day, one can see the effect aerosols have on absorbing and scattering light. Aerosols are small particles of various origins. Dust, sea salt, soot, sulphuric acid and ammonium sulphate are examples of important aerosols in the atmosphere. If humidity is high, they clump together to form tiny droplets. The role of aerosols in climate change is many-faceted and sometimes not well understood. A direct role is in their scattering of light. Indirectly they are important in cloud formation. In the stratosphere, aerosols cool the Earth, but in the troposphere they can act as greenhouse gases.

About half of the particulate matter in the atmosphere comes from anthropogenic activities, such as the burning of fossil fuels and biomass. Quantitative information on anthropogenic sources has been lacking and projections for the impact on future climate have up until recently only been made in very general terms. Aerosols might, for example, significantly counteract climate warming in the Northern Hemisphere. New quantitative results, presented in a 1992 update of the IPCC report, indicate that sulphate particles over the industrialized part of the world have actually veiled a large part of the greenhouse effect. This veiling effect would be most pronounced in sunny areas downwind from industrial regions, such as in the southeast part of the United States and in the Eastern Mediterranean area (Figure 3.13).

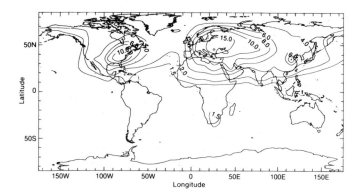

Fig 3.13. Simulated concentration of sulphate based on emissions

The aerosols have a short lifetime in the atmosphere, only a few days or weeks. A reduction in emissions from combustion and burning would thus lead to an immediate decrease in the atmospheric concentrations. Better air quality could thus, ironically, lead to a more pronounced greenhouse effect. Or rather, cleaning up the air could make it possible to statistically validate a warming trend that so far has been partly counterbalanced by dirt.

Natural phenomena such as volcanic eruptions can also add significant amounts of aerosols to the atmosphere, which in turn can counteract the greenhouse effect. A single eruption can cool the Earth by 0.1 to 0.2°C for one or two years. The question is whether volcanoes can also affect climate on a longer time scale. The total effect seems to be small, however, as the dust and other aerosols only stay in the atmosphere for 1 to 3 years, which is short enough for the oceans to buffer the climate system. If there was a period of sustained volcanic activity, it might delay the effects of greenhouse warming.

Another natural source of sulphate particles is from plankton in the form of dimethyl sulphate (DMS). This compound has an important role in cloud formation as it provides a nucleus around which humidity in the atmosphere can condense into a droplet. Scientists do not yet have a good quantitative understanding of how it affects the microphysics of the cloud, however. Neither do they know how a change in climate could affect the DMS-producing plankton and the concentration of the compound in the atmosphere. It has been suggested that a warmer climate might mean more DMS, thus increasing cloudiness.

Chapter 4

Adding up the numbers

CARBON DIOXIDE, METHANE, NITROUS OXIDE and CFCs exist in varying concentrations in the atmosphere and have varying rates of increase due to human activities. What does it all mean? What exactly is the relationship between these greenhouse gases and the future climate? In what direction should policies of reduction be directed?

These seemingly simple questions do not have any simple answers, but breaking them down before attempting to answer them in their entire complexity will be the first step in explaining both scientists' confidence in present knowledge and the limitations in their ability in making detailed predictions about the future. This and the following chapter will hopefully help explain why some scientists warn about placing too much confidence in models of climate change, while others do not hesitate to call for political actions to reduce emissions of greenhouse gases.

ENERGY BUDGETS

In its simplest terms, Earth's climate is determined by incoming solar radiation, the Earth's ability to reflect that radiation, and the ability of the atmosphere to trap heat. Using these variables, one can create a simple equation for flux of energy, which is well understood and can be calculated with some confidence.

A typical global average energy budget gives a natural greenhouse effect that increases the Earth's temperature by about 30°C above the temperature of an Earth without an atmosphere. Without the atmosphere, the global temperature would be about − 18°C. The most important greenhouse gas is water vapor, which accounts for 60 to 70 percent of the greenhouse effect in a clear sky, mid-latitude atmosphere. Carbon dioxide's contribution to the present greenhouse effect is about 25 percent. These contributions are called the climate forcing potential and are based on well understood physical properties of the greenhouse gases, such as their ability to absorb and reflect solar radiation in different parts of the spectrum. (Table 4.1)

Table 4.1   Radiative forcing relative to CO_2 per unit molecule change

Trace gas	ΔF for ΔC per molecule relative to CO_2
CO_2	1
CH_4	21
N_2O	206
CFC-11	12400
CFC-12	15800
CFC-113	15800
CFC-114	18300
CFC-115	14500
HCFC-22	10700
CCl_4	5720
CH_3CCl_3	2730
CF_3Br	16000
Possible CFC substitutes:	
HCFC-123	9940
HCFC-124	10800
HFC-125	13400
HFC-134a	9570
HCFC-141b	7710
HCFC-142b	10200
HFC-143a	7830
HFC-143a	7830
HFC-152a	6590

OVERLAPPING WINDOWS

Calculating the greenhouse effect is not as simple as adding the effect of each gas, however. The abilities of the various gases to absorb and reflect infrared radiation partly overlap. The situation can be compared to adding more glass to a greenhouse. If a window is added where none exists, it would be much more effective in trapping heat than if it was added as a second layer outside already existing windows. For example, the roles of methane and nitrous oxide are reduced by half due to their overlap in the absorption band with water vapor. The CFCs, on the other hand, fill in a window and thus becomes relatively more important.

In addition to the direct effect of the greenhouse gases, the warming of the Earth will increase the amount of water vapor in the lower atmosphere. This in turn enhances the greenhouse effect, as water vapor is in itself a greenhouse gas.

CONCENTRATION AND TIME

The ability to trap heat is also a function of the concentration of each gas in the atmosphere. Water vapor is in this respect the most important greenhouse gas followed by carbon dioxide. The other trace gases are present at much lower concentrations.

A somewhat more complicated picture emerges when taking changes in concentration into account. In figure 4.1 and 4.2, one can study the relative contribution of each gas based on trace gas concentrations over time. Water vapor is in this context considered to be constant and the focus is on the gases that have increased in the atmosphere. During the decade 1980 to 1990, 56 percent of the radiative

54

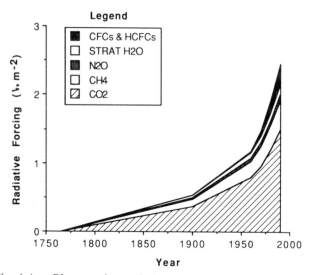

Fig 4.1. Changes in radiative forcing due to increases in greenhouse gas concentrations

forcing has been due to changes in carbon dioxide , 15 percent due to methane (directly and indirectly via increases in water vapor), 6 percent due to nitrous oxide and 24 percent from CFCs.

Regardless of future energy policies, carbon dioxide will remain the most important contributor to any increase in the greenhouse effect. If we continue with a "business-as-usual" scenario, carbon dioxide will account for more than 60 percent of the predicted climate warming. The HCFCs, which now contribute only marginally to the greenhouse effect, will in the next century become increasingly important.

A FOCUS ON EMISSIONS

Another method for comparing greenhouse gases is summarized in the term Global Warming Potential, which

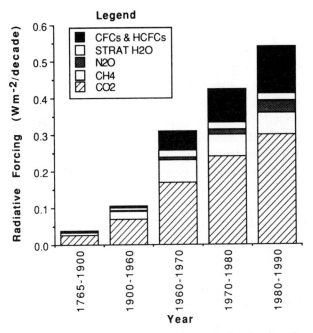

Fig 4.2. Decadal contribution to the radiative forcing due to increases in greenhouse gas concentrations

is a calculation that tries to bring in all the factors that can produce climate change over time. The emphasis is on emissions. This might be the most important concept from the policy point of view as it allows policy makers to compare the effect of cutting down various emissions. The concept of Global Warming Potential is, however, not as straightforward in physical terms as the previous ways of comparing greenhouse gases and several uncertainties enter the equations.

The most important variable is, of course, how large future emissions will be. A second difficulty is in accounting for the lifetime of the greenhouse gases in the atmosphere. Methane, for example, is not very stable and will break down to carbon dioxide and water. The direct effect of methane is thus fairly shortlived and a reduction of emissions will have immediate consequences for atmospheric concentrations. The CFCs, at the other extreme, are extremely stable and will stay in the atmosphere long after emissions have been cut. The calculated effect of each gas is thus not only dependent on its concentration and molecular forcing potential but on what time period one considers.

A third factor often entered into the Global Warming Potential number is the chemical changes in the atmosphere caused by a greenhouse gas. One can thus include the effect of excess water vapor in the stratosphere created by the oxidation of methane. Ozone chemistry in the troposphere and the effect of nitrogen oxides and other hydrocarbons than methane enter the equation here, but also introduce uncertainties as the chemistry is not always well understood in quantitative terms.

The effect of CFCs on stratospheric ozone has not yet been included in the calculation of global warming potential, only the direct greenhouse properties of these compounds. Another source of error is in using data from HCFC-22 as relevant for all HCFCs that may be used as substitutes for the long-lived CFCs.

The highest Global Warming Potentials can be found among the CFCs and related compounds. Then come nitrous oxide, methane and carbon dioxide in declining order if one considers the effect of each kilogram of gas that is released. Taking the magnitude of emissions into account, carbon dioxide remains the most important greenhouse gas in global warming (Table 4.2).

A conclusion that is not as evident from previous calculations is that indirect greenhouse warming may become significant. This is especially true for the production of water and carbon dioxide from methane. The

Table 4.2   Direct GWPs for 100 year time horizon

Gas	GWP	Known indirect component of the GWP
Carbon dioxide	1	None
Methane	11	Positive
Nitrous oxide	270	Uncertain
CFC-11	3400	Negative
CFC-12	7100	Negative
HCFC-22	1600	Negative
HFC-134a	1200	None

contributions of combustion gases such as carbon monoxide, nitrogen oxides and non-methane hydrocarbons may also become important. The magnitude of these effects may need further revisions, however, and in the most recent update of the Global Warming Potentials, the indirect effects were left out because the uncertainty was believed to be too great.

A second important point is that the CFC substitutes, with the exception of the relatively short-lived HCFC-123 and HFC 152a, will become troublesome greenhouse gases looking at a 20-year perspective. However, their Global Warming Potential decreases over time and in the long run they have a much lower warming potential than the long-lived CFCs they replace.

SOLAR INPUT

A change in the concentration of greenhouse gases is not the only way to change Earth's climate. In a geological time span, variations in the planet's orbit around the sun and the tilt of the polar axis are important in determining climate. Solar activity, as can be studied in sun spot activity, may also affect how much radiation reaches the Earth.

Could these factors explain a climate warming trend disregarding the effect of greenhouse gases? The question has been put many times, but these effects appear to be very small compared to that of changes in greenhouse forcing.

The influence of sun spots on climate has been brought up in relation to a discussion of an imminent cooling period corresponding to "little ice ages" in between glacial periods. The sun spots are not in themselves important but are thought to be an indication of solar activity. It has been difficult to prove any correlation between sun-spot activity and changes in past climates. Nevertheless, the Washington-based Marshall Institute has used sun spot data to suggest that we are on the way into another little ice age, which could counterbalance any future greenhouse warming. The IPCC Scientific Assessment agrees that such a change might occur at some time in the future but writes that the change would be so small that it would be swamped by the enhanced greenhouse effect. It might, however, be important to study such effects to be able to interpret temperature records correctly.

There are also variations in the solar diameter with an 80-year periodicity. The related changes in solar radiation have been studied by satellite and they could be sufficient to cause changes in the global mean temperature of about 0.1 to 0.2°C.

Solar irradiance can be studied by carbon-14 dating of tree rings as the carbon-14 production in the stratosphere is determined by solar activity. Using this method, one can see a statistically significant solar cycle of 200 years over the past 8,500 years.

CHANGING ORBITS

If the known climate effects of solar activity are very small, the opposite is true for variations in the Earth's orbit in relation to the sun. Changes in the shape of the orbit vary on time scales of 1,000 years or more and can change the solar input by up to 10 percent locally. Orbital changes on a 10,000–100,000 year time-scale were the major cause of the coming and going of ice ages during the Quaternary period.

Nevertheless, global climate changes due to orbital variations are small compared to the greenhouse effect. Variations in incoming solar radiation over the past 10,000 years show that the average change caused by greenhouse gases in one decade is 15 times that caused by orbital changes.

Chapter 5

A model world

W ITH THE LIMITED EFFECTS OF CHANGES IN solar radiation and effects of volcanoes on the time scales that are relevant for the current climate warming trends, atmospheric scientists have kept their focus on the climate forcing potential of the greenhouse gases. As outlined earlier, calculating radiative forcing potential for the various greenhouse gases is a fairly straightforward exercise. Translating these numbers to global temperature changes immediately adds more dimensions to the problem. One has to enter the world of atmospheric modelling — the art of re-creating Earth with its atmosphere and oceans within a computer.

The simplest climate models are based on changes in the flux of energy at the top of the troposphere. In this type of model, the globe is considered to be uniform in all directions and the mean global temperature is only dependent on changes in the flux of energy and on Earth's heat-storing capacity, for example on how changes in temperature may be moderated by the oceans.

This simple model assumes the climate to be the same all over the globe, which of course is a very simplified representation as is apparent to any visitor to polar or tropical regions. Neither does it take differences of altitude into account. The next necessary step is thus to add dimensionality to the model. One example is a model that takes into account the temperature differences between tropical regions and the poles. The atmosphere is heated at low latitudes and cooled at high latitudes and it is the difference in energy between these regions that drives atmospheric winds and oceanic currents. About half of the thermal energy is transported by the oceans and half by the atmosphere.

ICY REFLECTIONS

The energy balance model includes a feedback mechanism that influences how changes in radiative flux will affect global climate, the so-called ice-albedo feedback. Albedo describes the ability of a surface to reflect solar radiation. Calm water has low albedo, because of its dark surface. Snow and ice, on the other hand, reflect most of the incoming sun light.

In areas at the border between snow-covered and bare land or between ice-covered and free ocean, changes in albedo can be quite important. A decrease in temperature will increase the area covered by ice or snow, thus increasing the reflection of light. The more light that is reflected, the less heat is trapped. Or phrased another way, the more snow or ice, the less heat will be absorbed. This creates a positive feedback loop, which enhances the cooling effect. The opposite happens when the snow or ice melt. The water or soil can absorb more heat, in turn melting more snow and ice. These feedback loops are important in considering climate changes in the polar regions and have been entered into computer models.

HEIGHTS AND CLOUDS

Another model emphasizes the altitude differences in the atmospheric temperatures. An important feature of this approach is the ability to model the vertical energy distribution that moves moisture from the ground up into the atmosphere, creating clouds. It puts emphasis on feedback mechanisms involving cloud cover and cloud thickness.

The quality of clouds, their height, thickness and water content will influence the atmosphere's ability to trap and reflect solar radiation. For example, a change in average cloud cover is about twice as effective in reflecting incoming solar radiation as it would be in trapping heat. An increase in cloud cover would thus decrease the warming of the Earth and vice versa. More water in the clouds, as one would expect in a warmer world, would also have an effect on the radiation balance as these thicker clouds are very effective in reflecting away incoming solar radiation.

Cloudiness cannot be dealt with only in terms of averages, however. It is quite possible that increased average cloudiness would actually mean fewer but higher clouds. The net effect of all these changes depends on atmosphere dynamics that cannot be studied well in the models that brought these processes to the attention of the atmospheric scientists.

TAKING IN ALL THE DIMENSIONS

Adding differences in altitude and latitudinal spatial dimensions to the original model of global temperature change have been important steps in understanding global climate change, but they are not sufficient. One would really like to take in all possible parameters and then let the computer calculate what is in store. This is, however, not possible with present computer capacity and with present deficiencies in understanding some basic processes. In spite of the difficulties, a step in that direction has been taken with the development of general circulation models (Figure 5.1).

A general circulation model can be compared to the computer programs that are used by meteorologists all over the world to make weather forecasts. It divides the atmosphere into a mesh of points placed in a three-dimensional grid. A fourth variable is time. For each of

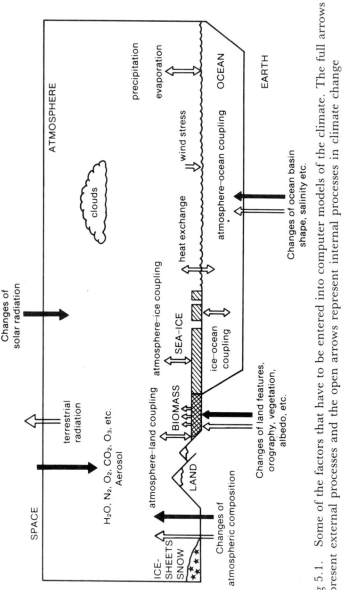

Fig 5.1. Some of the factors that have to be entered into computer models of the climate. The full arrows represent external processes and the open arrows represent internal processes in climate change

the grid points, the computer can calculate vertical transfer of solar radiation and heat, the transport of moisture and dry air, and water leaving the atmosphere in the form of rain or snow.

The atmosphere dynamics, in turn, depend on other physical, chemical and biological processes, for example how the ocean can absorb and transport heat, which also has to be entered into the computer calculations. Other such factors include interactions with land masses, which are determined by soil moisture, snow cover and ground temperature.

GLOBAL CONFIDENCE AND REGIONAL DOUBT

Anyone who is aware of the difficulties in making weather forecasts that hold for more than a few days, would intuitively doubt the ability of general circulation models to predict climate for decades to come. The wariness is both warranted and not. A reason to trust models is that weather is not the same as climate. Rather, climate can be thought of as average weather. In fact, climate statistics are gathered over a period that is much longer that what is predicted in normal weather forecasts. Usually, 30 years of data are used to define "normals". The "noise" in weather patterns in the models should thus be averaged out over time.

There are problems, however. The models that are used today do not deal adequately with how the atmosphere interacts with the oceans. Neither can cloudiness be modelled accurately as the grid system is much too coarse to catch all the processes that create clouds (Fig 5.2).

Despite the problems associated with computer models, atmospheric scientists place a significant amount of confidence in their ability to predict global climate changes. Comparisons between different models also give fairly

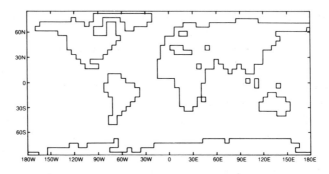

Fig 5.2. The land-sea mask for a typical climate model shows how coarse the grid system is

consistent results in predicting temperature increases in clear sky situations. For a doubling of carbon dioxide , current models show a range of temperature increase from 1.9°C to 5.2°C. Most results lie within the 3.5 to 4.5°C range.

Modelers are much more wary of drawing conclusions about regional climate. A further look into the deficiencies of the general circulation models helps explain some of the difficulties and needs for improvement.

OCEANS AND CLOUDS

The interplay between the oceans and the atmosphere is crucial in determining climate patterns. On one level, the oceans provide a sink for carbon dioxide, removing it from the atmosphere via physical and biological pumps as explained in the chapter on greenhouse gases. The oceans are also heat reservoirs with a dynamic transport system equal to that of the atmosphere. They have "weather" systems in themselves that are far less studied than the atmospheric ones. They include surface heating by the atmosphere, small-scale vertical mixing of the water and large-scale transport by ocean currents.

Early attempts to take into account the effect of the ocean did it in very simplified ways. For example, the oceans were treated as "swamps" or as a simple mixed layer, both of which neglect horizontal energy transport. The swamp model treats the oceans as a wet surface with no heat capacity, whereas the mixed layer models takes the heat storage capacity into account but considers the surface and deep water uniformly mixed. In the past few years, more realistic attempts to include the effects of the oceans in computer models have been actively pursued. Much research remains, however, before this problem of coupling the atmosphere and the ocean can be resolved. It will also require more powerful computers than are now available.

The formation of sea ice may create other serious problems in the general circulations models as this process provides a positive feedback loop in the model, where small deficiencies may amplify into large effects. Most models include the feedback mechanism, but not necessarily with due attention to physical realism.

Cloud formation in general is a problem in models, as future cloud properties could either make large increases or possibly large decreases in the net radiation budget of the troposphere. Generating scenarios for cloudiness is problematic, as cloud properties are determined by processes that occur on a much smaller scale than the general circulation models work on. An apparently even cloud cover has wide-spread variation on the 100 kilometer scale used in the models and some clouds can be less than a kilometer wide. The same problems hold for the thickness of clouds.

In addition to these difficulties, the ability of clouds to reflect and absorb radiation is not well enough understood for atmospheric scientists to enter a potential change in

70

properties into their models. A comparison of 17 different models illustrates the difficulties in modelling cloud interactions. Whereas the model results were rather consistent when assuming clear sky conditions, their predictions had a threefold variation when cloud-sensitive parameters were added (Figure 5.3).

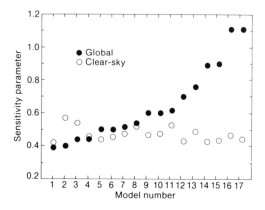

Fig 5.3. Seventeen different models predicted how sensitive the climate is to changes in radiative forcing. Assuming clear skies the results are similar, while different treatment of clouds in the various model results in very different climate sensitivity

Just as clouds reflect and absorb solar radiation, so does the surface of the Earth. So far surface albedo has been treated in an oversimplified manner in the models, sometimes only differentiating between snow-covered and bare surfaces.

Another simplified factor is how precipitation affects soil moisture. Most models deal with the soil as a bucket, which is filled up as it rains or snows. As the bucket fills up, the overflow is added to the surface runoff in streams and rivers. Part of the water in the full bucket is also assumed to evaporate directly back into the atmosphere. In reality, a slow drizzle and a downpour have quite different effects on soil moisture. This difference could determine whether a farmer has a well-watered crop or, conversely, whether much of his soil ends up in the river and his land is subject to drought.

A SYSTEM IN TRANSITION

The final problem of climate models is of a different nature. It deals with if one should consider the Earth as a system in equilibrium or a system in change. As the concentrations of carbon dioxide and other greenhouse gases are growing in the atmosphere, it would be appropriate to base the calculations on a continuously changing system. This should include such parameters as the lag time for increased temperature as part of the heat is stored in the oceans (Figure 5.4).

Calculations based on continuous change have only been made to a limited extent so far, as they require better

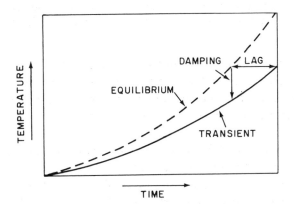

Fig 5.4.   Equilibrium and transient warming

ocean circulation models as well as much more computing capacity than is readily available. Instead modelers base their calculations on the assumption that the system has reached an equilibrium at some increased concentration of carbon dioxide. Most often, one looks at the effect of increases in greenhouse gases by comparing model simulations with present concentrations in the atmosphere to those with a doubling of the concentrations.

The difference between looking at equilibrium and taking into account transient change is not only one of time lag or overall magnitude of the change at a certain point in time. It also affects the regional distribution of the climate change. Recently, model studies of transient change have improved. The results confirm previous findings on a global scale but show that the warming over the northern North Atlantic and the southern oceans near Antarctica is much less than in the equilibrium models.

VALIDATING MODELS

Taking all the uncertainties in the computer models that are used for climate predictions, one can wonder how atmospheric scientists dare to say anything at all. The answer lies in the extensive work that has been done to validate the models — to run them with known variables and compare the results to present or past climates. This kind of validation should reveal systematic errors that are common to many models and that would not be caught in a simple comparison between models.

Two of the climate parameters studied are air pressure at sea level and temperature. Air pressure is connected to the paths of storm systems and is thus very important in the prediction of weather patterns. The models can realistically simulate the major storm tracks in mid latitudes, but make important errors on the regional scale. The pressure patterns are important in determining wind directions and most models manage to show the broad features of zonal winds realistically. Only a few more recent models manage well with the path and intensity of the jet stream, however.

Temperature patterns are successfully simulated by the models, except for some regions (Figure 5.5). Eastern Asia gets too cold and Antarctica too warm in the models compared to what we know of present day climate. The models also manage well in simulating the changes in temperature over the year. This is important for validation, as a seasonal cycle includes large changes in the atmosphere-ocean system, which has been considered a weak point in modelling.

Precipitation is another parameter in which the models manage well on the global scale but err on regional patterns.

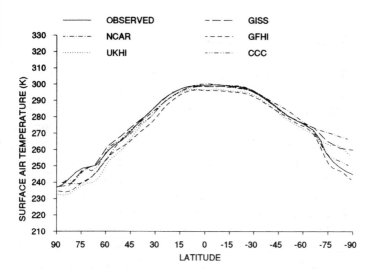

Fig 5.5. Zonally averaged surface temperatures in the winter for various models and as observed

They also have difficulties predicting the seasonal variation in precipitation and give more rain and snow in winter than the real weather does at present.

DRIER OR WETTER SOIL?

The problems with modelling regional temperature and precipitation are accentuated when studying more complex phenomena such as ground moisture. Soil moisture may be one of the most important factors for the success of agriculture and the maintenance of current natural ecosystems in the future, but here the models show large differences (Figure 5.6). A lack of global soil moisture data makes it difficult to validate the models against present reality. Even if the data available show that the models reflect most large scale characteristics, the representation

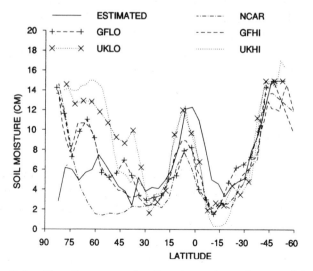

Fig 5.6. Zonally averaged soil moisture in various models in June–July–August and as observed in July

and validation of soil moisture in current models are still relatively crude.

Soil moisture may depend on feedback loops where large decreases in soil moisture give higher surface temperatures (the lack of cooling from evaporating water) and less cloudiness leading to less precipitation to moisten the soil. Other feedback loops that can create problems in modelling future climate are the treatment of snow cover, of sea ice and the ability of clouds to reflect radiation. The models manage quite well with snow cover on the large scale and this should not distort predictions of the future global climate. Improvements are needed before the models can predict snow cover on a regional scale, however. The treatment of sea ice is much more problematic as there are not many data on present conditions to compare with, just as in the case with soil moisture.

To validate the models' treatment of the ability of clouds to reflect radiation, atmospheric scientists have compared model results to the amount of heat leaving the top of the atmosphere. There is a nice fit of the curves from present-day data and the results predicted by the models in studying how this parameter changes over latitudes. The problem is in modelling the albedo close to the poles.

EXTREME WEATHER

Average climate in a region or on a global scale is one thing but equally important in how climate actually affects people are extreme climatic situations, such as long droughts and storm events. An example of the latter is weather associated with the El Niño Southern Oscillation, which is coupled to high sea-surface temperatures in the eastern tropical Pacific. Here the models do quite well in simulating the changes in atmospheric circulation at least for intense

El Niño episodes. The drought in sub-Saharan Africa during most of the 1970s and 80s can also be seen in the models from year to year.

A recurring event with changing intensities and timing is the summer monsoon. It is important for the people who rely on it for agriculture as an early arrival or delay easily can cause flooding or drought. It is also interesting for validating models as it involves abrupt changes in atmospheric circulation patterns and it would give an indication of how well the models do on a regional scale. The models can simulate the gross features of summer monsoon precipitation, but there are some significant deficiencies.

The El Niño, the Sahelian drought and the onset of the summer monsoons deal with regional scale changes in weather. Some extreme weather, such as tropical storms, is much more localized. Simulating these events is still beyond what the models can perform. The extreme weather that can be predicted on purely statistical grounds is that there will be an increase in frequency in high temperatures.

LOOKING AT THE PAST

A quite different approach to model validation is to compare the results with past climatic changes. The most important period used in these comparisons is the time since the last glaciation 18,000 years ago. An example of these comparisons is a map of the distribution of spruce pollen, which can be directly measured in the old pollen records. The area covered by spruce is determined by temperature and precipitation, which can be simulated with the models. The maps of simulated spruce distribution and observed pollen overlap quite nicely (Figure 5.7).

Other palaeo-climatic studies of lake sediment, ice and soil cores show that there are general agreements between

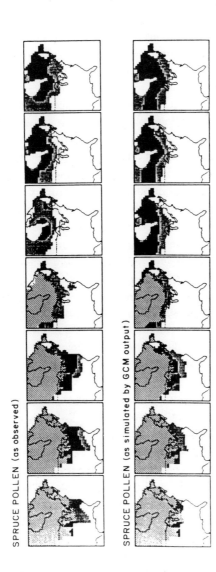

SPRUCE POLLEN (as observed)

SPRUCE POLLEN (as simulated by GCM output)

18 kbp   15 kbp   12 kbp   9 kbp   6 kbp   3 kbp   0 kbp

Fig 5.7. Observed (upper row) and simulated (lower row) percentage of spruce pollen for each 3000-year interval from 18 000 years ago to present

models of past climate and what it was like according to remnants of pollen, plankton and other biological clues to the climate. But, just as with comparisons with present climate, palaeo-climatic studies reveal that the models are not always as accurate in simulating regional climatic patterns.

Looking back even further in time, there is an interesting period about 125,000 years ago, in the previous interglacial. There is some evidence of warmer conditions, especially in high latitudes. The carbon dioxide levels were above pre-industrial levels and this period would thus provide a parallel to a future greenhouse world. Unfortunately, there are not enough palaeo-climatic data to use this period for validating models.

A CONCLUSION OF CONFIDENCE

Adding up the validations against past and present climates, the IPCC authors conclude that general circulation models used today can simulate climate "sufficiently close to reality to inspire some confidence in their ability to predict the broad features of a doubled carbon dioxide climate at equilibrium, provided the changes in sea surface temperature and sea ice are correct". There is as yet no guarantee that they are correct, but comparison with past climate makes it likely.

Chapter 6

Model results
—the future climate

DESPITE THE INADEQUACIES OF THE PRESENT general circulation models of the atmosphere, these are the only tools available to scientists to make predictions about the future climate. And if we want to create options for the future, it is necessary to use the tools available. The alternative would be a wait-and-see attitude, which would limit our possibilities of adjusting.

Atmospheric scientists have thus placed great effort into interpreting model results to be able to create scenarios for the future climate. In the IPCC Scientific Assessment more than 20 simulations by nine modelling groups have been assessed in the latest up-to-date review of what might be in store for a future with doubled levels of greenhouse gases in the atmosphere, expressed as carbon dioxide equivalents. A doubling is used as a convenient benchmark, which makes it easier to compare model results. With a business-as-usual scenario for emissions of greenhouse gases this would occur around the year 2025.

HIGHER TEMPERATURES

The models all agree that the global mean temperature will rise. The range of temperature increase is 1.9–5.2°C. Most results lie between 3.5 and 4.5°C, but the range of uncertainty can really not be decreased as long as clouds cannot be better modelled. Attempts to narrow the range of uncertainty by comparing model results to present-day temperatures are also problematic as effects of aerosols, stratospheric ozone depletion, and clouds may have suppressed the expected warming.

The warming is not evenly spread over the globe. All the models predict an enhanced warming in high latitudes in late autumn and winter (Figure 6.1). This is caused by feedback processes involving the sea ice. In the summer,

82

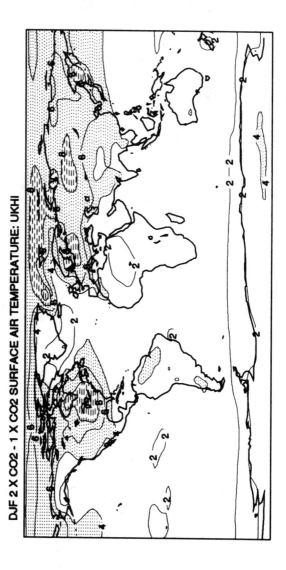

Fig 6.1. One example of model results showing the surface air temperature due to doubling of carbon dioxide for December, January and February

climate warming in the polar regions is less pronounced. The sea is kept cold by melting ice, which moderates the temperature.

In the northern mid-latitudes, warming in the summer may be enhanced because the land surface is too dry to allow for evaporative cooling. This in turn can lead to fewer low clouds, which further enhances surface warming (Fig 6.2).

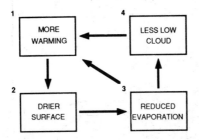

Fig 6.2. Soil moisture temperature feedback through changes in evaporation and low clouds

In the tropics, the warming will be less than the global average. The average warming is typically 2–3°C. The excess heat is used to evaporate soil moisture rather than heating the air close to the surface of the Earth. The upper atmosphere is, on the other hand, heated more than the global average.

SPEEDING UP THE WATER CYCLE

As indicated by the soil-moisture feedback, water evaporates more quickly in a warmer world. On a global scale this leads to more precipitation as the amount of water in the system is fairly constant. It is a general speeding up of the global water cycle. The models predict global increases ranging from 3 to 15 percent.

As with temperature, the <u>precipitation changes are unevenly distributed</u>. High latitudes and the tropics will receive more precipitation throughout the year and the mid-latitudes more rain in the winter. The high resolution models give an increase in precipitation by 10–20 percent averaged over land between 35 and 55°N. This is a belt with a southern boundary along the Mediterranean Sea, central China and central United States and a northern boundary along the southern Baltic Sea via the northern Mongolian border to the Hudson Bay in Canada. How the precipitation will be distributed within this belt depends on the topography on a scale that is much smaller than what the models can work on.

There has been a lot of speculation on whether climate warming will lead to less rain in areas now hit by devastating periodic droughts, such as in the Sahel region of Africa. There does not seem to be any evidence for that in the models, however. They do not predict any large changes in precipitation in the dry subtropics.

In some areas, too much rain is more of a problem than too little, examples being the wet subtropics and regions where precipitation is governed by the South East Asian monsoon. Here, the models' results are difficult to interpret. Most models give an increase in the strength of the monsoon with heavier rainfall, but the processes that determine precipitation are not very well resolved in the model grid. The changes are also small compared to natural variation. A further complication is that the monsoon might shift geographically.

SOIL MOISTURE

Will the land become drier or wetter with the predicted changes in temperature and rainfall? The answer depends on both these primary climatic variables and on what happens with the water once it has hit the ground. This includes the effects of soil type and vegetation. The models take surface hydrology into account in a simple way, but neglect any direct effect of carbon dioxide, which could enhance the vegetation's ability to use soil moisture.

The major results in the different models, and a result on which they agree, is that the soil will be wetter in the winter on the continents in the northern high latitudes. There will be more rain than now, more melt water from snow, and temperatures will not be high enough to dry the soil by evaporation.

In the summer, one can see the opposite trend. Higher temperatures allow for more evaporation leaving less moisture in the ground. This will be important for agriculture and forestry as will be discussed in later chapters. Some high-resolution models give reductions in soil moisture of 17–23 percent over the 35 to 55°CN latitude band, but the confidence in these results is low. The general picture is that the ground will start drying up earlier in the spring as a warmer climate will bring an earlier snow melt than at present. How fast the ground dries depends on how soil moisture interplays with cloud formation and precipitation. It also depends on how saturated the soil is with water in the winter. The more water it can store from the winter precipitation, the longer the moisture will last as temperatures increase in the spring and summer.

HOW CAN THE SEA AND ICE CHANGE THE CLIMATE?

Going further north or far south toward the polar regions, the extent of sea ice is a major determinant of climate. With a doubling of carbon dioxide, the sea ice is significantly reduced, in some models it even disappears completely in the summer.

The ice-free sea also affects weather patterns. If, for example, ocean currents change, the interplay between the sea and the atmosphere could become quite different on a regional scale. This in turn would affect the paths of low-pressure systems. Two features that the climatologist would like to predict are therefore changes in the large-scale atmospheric circulation at sea level and changes in the deep ocean circulation. To simulate changes in deep sea currents, one has to rely on atmospheric models that have been coupled to dynamical models of the sea and there are very few experiments of this kind that have been done. The two main features that have emerged are that the large warming in high latitudes will be distributed in the deep ocean, and that there will be a weakening of the north-south oceanic circulation.

A MORE EXTREME CLIMATE?

Will the climate vary more from day to day, or from year to year, with global warming? Is there a greater risk of extreme weather? These questions can be answered in two different ways illustrated by the shift or spread of a temperature-distribution curve (Figure 6.3). The range of "normal" temperatures can be illustrated with a bell-shaped curve of a certain height and width. The change in variability of the temperature is then represented by a wider curve, where there would be more extremely hot and more extremely cold days. The whole curve could also shift without changing its shape. With a shift toward warmer temperatures, there would indeed be more extremely hot days. There would on the other hand be fewer extremely cold days. However, so-called "blocking situations", such

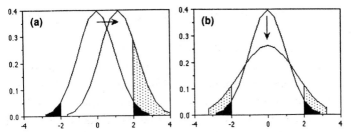

Fig 6.3. The curves show the difference between (a) only increasing the mean temperature and (b) increasing the variability without changing the mean. In both cases, the result is more extremely warm days.

as persistent high pressure systems, may become more frequent. This could lead to more severe droughts in the mid-latitudes, for example in Europe.

Whether the climate will actually vary more than today is much more difficult to assess. One reason is the fact that one needs a larger statistical base to be able to calculate this kind of change and there is a lack of model simulations providing that information. What can be said is that the day-to-day variation seems to be less in a warmer-world scenario than is the case today in some areas. Looking at the year-to-year variation, there are no meaningful patterns except around the sea–ice margin.

It is even more difficult to assess changes in the variability of precipitation as the natural variability is higher than for temperatures. The only consistent change is an increase in convective precipitation, which is local in nature. There might be more intense local rain storms at the expense of gentler but more persistent rainfall.

STORMY PATTERNS

Will the winds change? Will storms be more severe? The models can only give an indication of the likely changes as storm tracks and hurricanes in particular are phenomena on a smaller scale than most models can resolve. Nevertheless, the IPCC authors give some pictures. At mid-latitudes, there may be a decrease in day-to-day and year-to-year variation in windiness — the winds will be more even as the low-pressure sysems will be less intense.

In the tropics, winds are determined by quite different mechanisms and hurricanes and typhoons may become more frequent. The changes are associated with increases in sea-surface temperatures and the El Niño/Southern Oscillation events. There are also suggestions that the intensity and thus the destructive power of tropical storms can increase when the sea surface temperature is above 27°C.

Another way to predict the frequency of storm events in the tropics is to use models with well-defined criteria for storm formation, for example sea-level air pressure below a certain number. One such attempt, with a prescribed increase in cloudiness, gave an increase of 20 percent in the number of storm days with a doubling of carbon dioxide levels. In another experiment where cloudiness was allowed to change, the number of storm days decreased instead. Summarizing the results, one can say that the maximum intensity of tropical storms may increase, but the distribution and frequency of the storms will depend on changes in the circulation in the tropics. These changes are not well simulated in the current models.

REGIONAL CHANGES IN CLIMATE

What policy makers world wide really want to know and what scientists so far have not been able to answer is how climate will change on a regional scale. What will it be like in south east Asia in the year 2025? Will the Sahel get hit by further droughts? What will happen in Australia? The problem, as outlined in the chapter on model validation, is that the simulation results are not very consistent on the regional scale. The spatial resolution is not good enough to make predictions for areas smaller than about 1,000 km square. For example, even if atmospheric scientists are confident in their prediction that precipitation will increase around the latitude 60° North averaged around the globe, it is impossible to see where, on the longitudinal scale, the rain will fall. The patterns are very much influenced by mountain ridges and other features of the surface topography, which are too small to be included in the models.

In spite of all these uncertainties, the IPCC Scientific Assessment has made an attempt at making a few regional scenarios based on high resolution models. The conditions used are a doubling of greenhouse gases according to the business-as-usual scenario for emissions and a best guess of the magnitude of climate warming at 2.5°C. Furthermore, there are assumptions that the pattern of equilibrium and transient changes in climate are similar and that regional changes in temperature, precipitation and soil moisture are proportional to global mean changes in surface temperature. The authors stress that the confidence in the estimates is low, especially for changes in precipitation and soil moisture.

The results from these exploratory studies are the following:

In central North America, the mean temperature will increase by 2–4°C in the winter and by 2–3°C in the summer. Precipitation increases are in a range of 0–15 percent in the winter with decreases of 5–10 percent in the summer. There will be 15–20 percent less soil moisture in the summer compared to today.

In south east Asia, the warming varies from 1 to 2°C throughout the year. The precipitation changes are limited to the summer, where they result in soil moisture increases of 5–10 percent.

In the Sahel region, similar temperature trends can be seen. The precipitation and soil moisture changes vary throughout the region but there might be a marginal decrease in soil moisture in the summer.

In southern Europe, the warming trend is somewhat stronger, 2°C in the summer and at least 2–3°C in the winter. It may rain more in the winter but summers will be drier. Soil moisture may drop by 15–25 percent.

In Australia, warming ranges from 1 to 2°C in the summer and is about 2°C in the winter. It will probably rain more in the summers, but it is impossible to interpret what that means for soil moisture as the models do not produce consistent estimates.

Regional climate will not only be affected by changes in global average temperature. Equally important may be regional changes in vegetation that are not directly related to the greenhouse effect. An illustrative example is the Amazon basin, where the tropical forest plays a significant role in determining rainfall. About half of all the rain is derived from evaporation in the area and deforestation would definitely affect the regional water balance and reduce rainfall. A complete deforestation of South America would lead to an irreversible decline in rainfall and vegetation cover over at least part of the region.

Chapter 7

Sea level

SITTING ON A ROCK ON THE WEST COAST OF Sweden, I can watch the waves slowly caress the shoreline. The water is lower than usual. There is a high pressure system stationed over Scandinavia. The sun is shining and the rays reflect in the water. This place is not always as peaceful. Deep low pressure systems push the water ashore and the waves beat the shoreline rather than caressing it. This is an area without noticeable tides, but the sea level can vary more than a meter.

I think of changes—of times past and times to come. Just a few hundred years ago, the rock I am sitting on was under water. The land here is still rising in response to recession of the inland ice during the last ice age. A hundred years in the future, the rock might be covered by water again as the sea rises in response to climate warming.

In a global perspective, sea-level rise has been a major concern in looking at the effects of climate warming. My tranquil contemplation on the shore of the North Sea is very far from the reality facing many people living on low islands or in coastal areas by river deltas. In those places, the sea may very well displace millions of people, who have to find new places to live and new agricultural land to cultivate, if such places and land exist.

MEASURING CHANGES IN THE PAST

Sea-level rise is not a new phenomenon. Much scientific attention has thus been focused on previous changes in sea level and understanding the various processes that have contributed is necessary to be able to make predictions about the future.

Measuring global mean sea level is far from simple, but estimates of sea-level rise from all sources add up to an average rise of 1–2 mm per year (Figure 7.1). The

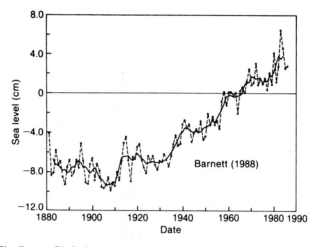

Fig 7.1. Global mean sea-level rise over the past century

difficulties in studying mean sea level lie in the fact that the land moves as well. In Scandinavia, for example, land is still rising after depression from heavy inland ice during the last ice age. The sea thus appears to sink. In other coastal areas, erosion eats away the coast line. The reverse process, the building up of silt from rivers, is no less important.

A major cause for the sea-level rise that has been recorded in the past 100 years is the thermal expansion of water in response to a warming of the atmosphere. In addition, there is more water in liquid form today than a century ago, as many small glaciers and ice caps are slowly melting and have been doing so since the Little Ice Age (approximately 1400–1800) (Figure 7.2). Together, these two processes have amounted to approximately 10 cm in the past century, about half from each source.

GREENLAND AND ANTARCTICA

The Greenland and the Antarctic ice sheets pose quite

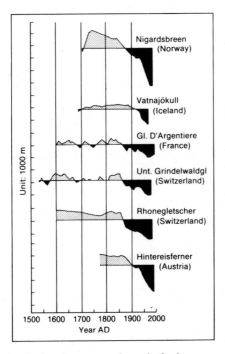

Fig 7.2. Variation in some selected glaciers as measured by their length

different problems compared to the small glaciers. These giant inland ice masses are not only affected by the climate but also affect it themselves and climate warming does not necessarily lead to less water captured in the ice than at present. Looking more closely at Greenland, it is difficult from present data to judge whether water-removing processes, such as calving of icebergs, or water-adding processes, such as the accumulation of snow, are ahead. This uncertainty also leads to difficulties in estimating the sensitivity of the Greenland ice sheet to increases in temperature. The contribution of the Greenland ice sheet to sea-level rise in the past century seems to have been somewhat less than for thermal expansion and from small glaciers.

The Antarctic ice sheet may still be adjusting to changes since the last ice age, making it difficult to calculate a mass balance of the water contained on this continent. There may also be regular changes in the West Antarctic that are unrelated to climate change but still influence sea level. Looking into the future, computer models simulating climate warming indicate that precipitation will increase close to the poles. This could mean an increase in the water captured in the ice, which would contribute to a lower sea level.

AN ANTARCTIC COLLAPSE?

There have been voices of concern that the West Antarctic ice sheet might collapse if the global temperature rose to a certain level and thus cause a massive release of ice into the oceans. The ice here has been pushed off the land and rests instead on the ocean floor. Here it serves as a barrier against the calving of icebergs from the continental ice sheet and is thus very important in determining the dynamics of gains and losses to the Antarctic ice. As it is grounded below the sea surface, it would also be sensitive to a rise in sea level.

Several researchers have tried to model the processes that could cause a collapse of the West Antarctic ice sheet. The models show that a collapse would require unrealistically large ice-thinning rates. A more likely scenario is that the West Antarctic ice sheet would only contribute an additional 0.1 mm per year to the sea level in the coming decades. Fast-flowing ice streams are not included in those models, however, and the possibility for fast changes in the dynamics of the ice streams show that the West Antarctic ice sheet

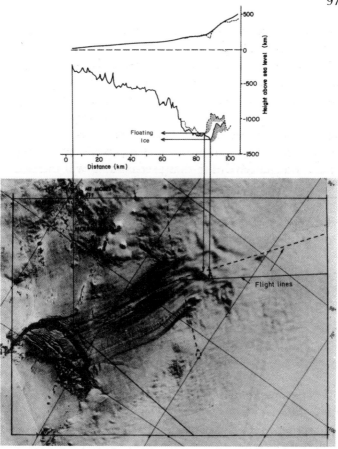

Fig 7.3. Fast-flowing ice streams, such as this one, might indicate instability on the West Antarctic ice sheet

can react quickly to a changing climate (Figure 7.3). Lack of oceanographic knowledge also adds uncertainty. Questions that need answers include how the circumpolar deep water will warm in response to climate warming and if circulation under the ice shelves will change.

Disregarding these unknowns, a typical greenhouse scenario shows that the West Antarctic contribution to sea level would be a decrease of 10 cm after 100 years as the accumulation of snow would still dominate. After 200 years, there would be an addition of 40 cm to the sea level.

A HIGHER SEA

Many scientists have tried to project future sea-level rise with various calculations by adding up the numbers for thermal expansion of sea water and the melting of water bound in ice. Their results range from + 10 cm to + 30 cm in the next 40 years (Figure 7.4 and 7.5). This is a faster rate than we have experienced in the past century. Some studies look further into the future. For a business-as-usual scenario, one estimate is a range from 20 to 70 cm rise in sea level by the year 2070.

What would happen if strict control policies reduced emissions of greenhouse gases? Not as much as one would think. The oceans and the ice sheets react slowly to changes

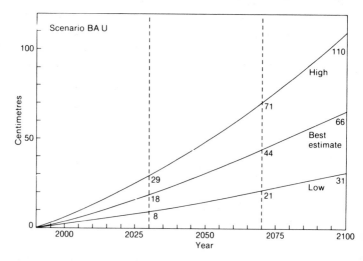

Fig 7.4. Global sea-level rise 1990–2100 for a business-as-usual scenario

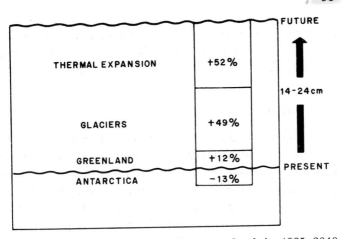

Fig 7.5.   Factors contributing to future sea-level rise 1985–2040

and the sea level would continue to rise long after emissions were cut. For example, if the increase in greenhouse forcing is completely stopped around the year 2030, the sea level will continue to rise at about the same rate for another century (Figure 7.6). A prompt decision would make a difference, however. If the emission levels were cut immediately, as in the strictest control scenario in the IPCC Scientific Assessment, the total sea-level rise in the year 2070 would only be about one third of that of the business-as-usual scenario.

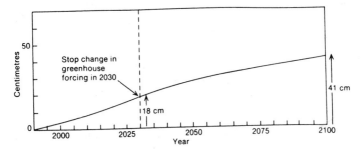

Fig 7.6.   The sea level will continue to rise for several decades after climate forcing is stabilized

REGIONAL EFFECTS

Predictions of average global sea-level rises may not mean much when trying to plan locally for future changes. In some areas, the land moves in ways that will compensate for the changes in sea level. In erosion-prone areas, on the other hand, the changes might be accentuated by beaches being washed into the sea. Changes in ocean currents and atmospheric circulation will also give pronounced regional differences. According to dynamic ocean models, one can expect a doubling or halving of the effect regionally as compared to the global mean sea-level rise.

In spite of the difficulties in giving exact numbers for future sea levels in different regions, it is possible to describe some of the problems that people in coastal areas will have to face. One should, however, remember that the pictures painted often use sea-level rises that are higher than expected in the coming 70-80 years. The pictures are nevertheless gloomy, especially for people living in low-lying delta regions. A one-meter rise in sea level would inundate 12–15 percent of Egypt's arable land and 17 percent of Bangladesh. In Bangladesh, about ten million people live on land that currently lies within one meter of sea level. Aside from the Ganges and the Nile, the list of vulnerable deltas is long: the Yangtze and Hwang Ho in China, the Mekong in Vietnam, the Irrawaddy in Burma, the Indus in Pakistan, the Niger in Nigeria, the Parana, Magdalena, Orinoco and Amazon in South America, the Mississippi in the United States and the Po in Italy.

The effect of inundation is at least twofold. Firstly, people will have to move, and there is not always somewhere else to go. Secondly, food production will decrease as arable land disappears under water or is destroyed by salination or floods. One calculation shows that 20 percent of the agricultural production in Bangladesh could be lost by a one-meter sea-level rise. The corresponding figure for Egypt is 15 percent. Fresh water wells may also become unusable (Figure 7.7).

Small islands will have to face the sea in completely new

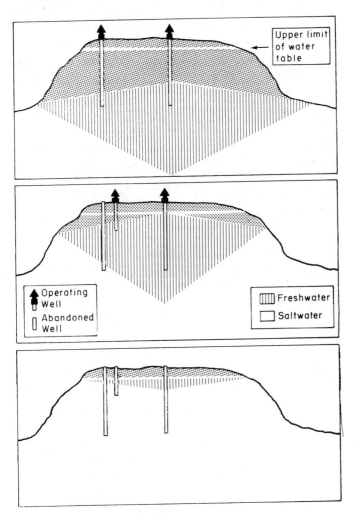

Fig 7.7.   Impact of sea-level rise on groundwater tables

ways. Coral atoll islands, such as Kiribati, the Marshall Islands, Tokelau and Tuvalu, Cocos, and Keeling Islands, are generally not more than three meters high. The 1,190 small islands making up the Republic of the Maldives also fall into this category. Their problem will be compounded when the coral reefs that act as natural wave breakers become buried by the sea. Unless the coral growth can keep up with the sea-level rise, the islands will have much less protection in severe storms than they have had so far.

The inhabitants of the small islands do not have the choice of moving upland, and might instead have to invest large sums of money in protecting their shorelines against the rising sea and erosion. The annual cost of protecting areas that are populated with more than 10 people per square kilometer may be more than 5 percent of the gross national product. This does not include costs for protection of ecologically important wetlands or problems with salt-water intrusion.

In some coastal areas, the flooding may be accentuated by changes in coastal storm patterns and the risk of more severe weather. The southeastern United States, the Indian subcontinent and the western Caribbean islands will all have to count on a higher base from which the storm surges build and hit the coast during hurricanes. An area that is today flooded by half-a-meter every twenty years will have a flood of one meter at the same time interval if the sea level is half-a-meter higher. In south Australia, there are areas that today are flooded every 100 years that might well get hit every other year by the same kind of floods given a half-a-meter increase in sea level. Calculations from Japan show that a one-meter sea level rise would threaten an area of 1,700 square kilometers where 4 million people live.

The cost of protecting the coasts can become quite high and it will be dependent on the economic development in an area whether it is feasible or even affordable to erect dikes, for example. Two of the countries for which calculations have been made about costs for protective measures are the United States and Japan. In the United States, it would cost about 100 billion US dollars to protect tourist resorts and vulnerable urban areas. A Japanese calculation for a one-meter sea-level rise gives a cost of 20 billion US dollars.

THREATENED WETLANDS

Coastlines are often highly exploited and populated areas. But equally important in discussing climate warming is the effect on natural ecosystems along the coastline, such as wetlands and estuaries. Attention to the role of these areas has increased over the past decade. Areas that were once considered wasteland are now often given an economic value. For example, salt marshes are important nursing grounds for many of the commercially important fishes. They also serve as a buffer in case of floods and storms and as cleaning filters for many pollutants that would otherwise severely affect the coastal waters. Some US researchers have calculated the economic worth of wetlands to be more than 10,000 US dollars per hectare.

The short-term effect of inundating sea water on the marshes might very well be a positive one, at least for fish production. The flooding releases nutrients on which the fish thrive and today one can actually see this trend in southeastern United States. In Louisiana, the juvenile shrimps seem to have increased as more marsh area provides breeding ground as the land floods.

The effects in the long run depend to a large extent on the rate of sea-level rise and on if the natural ecosystems can keep up with the pace by moving upland (Figure 7.8). However, there is nothing that indicates that wetlands

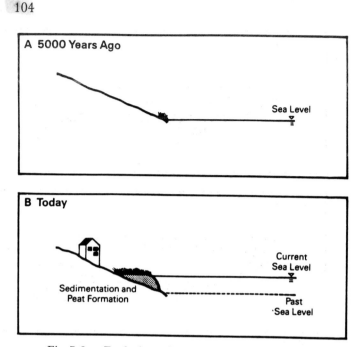

Fig 7.8.   Evolution of a marsh as sea-level rises

would be able to keep up with a rise in sea level of 1 cm per year. The problem is accentuated in areas where wetlands have to compete with human exploitation of the land, which will often be the case where buildings and other structures have been erected at a previously safe level from the sea and marshy coastal areas.

THE FISHING INDUSTRY

From an economic point of view, many coastal areas will also be affected by changes in the fishing industry. Fish catches in a region are determined by anthropogenic factors, such as the intensity of fishing, but also by global-scale

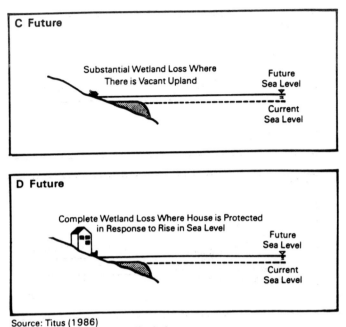

Source: Titus (1986)

Fig 7.8. cont.

natural phenomena, such as ocean currents that determine the availability of nutrients. In a global-warming perspective, there may well be changes in ocean currents and upwelling zones along the coasts that can affect fish production. The distribution of warm and cold water in the polar regions will also play a role in where various fish populations may increase, decrease or migrate. In general, the sea fauna is fairly adaptable but the economy in a specific region can of course be affected if catches increase or decrease. This is especially true in nations where the fisheries form an important and narrow base for the economy as in some coastal countries near upwelling regions. Examples of these are Mauritania, Namibia, Peru and Somalia.

Chapter 8

Agriculture

WILL NEXT YEAR'S CROP BE ANOTHER FAILURE or will it be possible to once again fill the stores? Will we have enough to eat? Will prices be high enough to be able to pay off the investments in new farm machinery? When are the rains coming? Will we have a late or early frost?

The chance of a farmer getting a good crop in a particular year to a large extent depends on the weather. A major question in discussing global warming has thus been how a new climate can affect farming conditions in different regions. This information is important as farming is an area where it in many cases is possible to adapt to new limitations.

Farming is also an economic activity in a complex international economic system, where the market price might be as important as the size of the harvest for an individual farmer and for the consumer buying the goods. Looking at it from the larger perspective, questions of global food security and regional self-sufficiency or dependency come into focus.

Many attempts at making scenarios for the future of agriculture have been made in the wake of climate warming, but the lack of data on regional climatic differences has made it difficult to make any predictions. Technological and social changes are even more problematic to foresee, as a look back on farming and global food production 50 years ago would make apparent. The focus in many reports has rather been to describe how sensitive different crops and agricultural systems are to variations in climate parameters such as temperature, precipitation and soil moisture. Attention has also been paid to the direct effect of increased carbon dioxide in the atmosphere as this gas acts almost like a fertilizer for some plants (Figure 8.1).

An offshoot of the climate-sensitivity studies is a look at present and future geographical ranges for different crops.

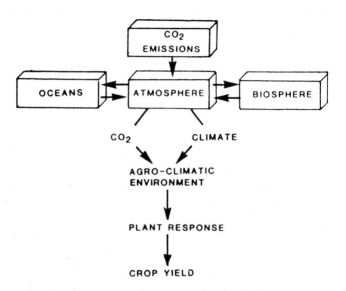

Fig 8.1.   Crop impact analysis

Will it be possible to grow maize further north, for example? How far south is it economically reasonable to grow wheat without irrigation? What other factors than climate would hinder a movement north of a crop as frost risks diminish?

What has as yet not been taken into account in future scenarios is the ability of farmers to adjust to changes by changing crop varieties or farm practices, even if attempts have been made in some of the response strategies to climate warming.

This chapter will start with the physiology of individual plants and then go up in scale in an attempt to point at some of the opportunities and limitations future farmers will face. Will there be enough rain to grow maize or should farmers grow wheat instead? Maybe it is time to invest in an irrigation system?

CARBON DIOXIDE AS A FERTILIZER

The very base for agriculture is the ability of plants to use solar energy to capture carbon dioxide from the atmosphere. This carbon is then bound in energy-rich compounds, some of which are eventually stored, for example as starch and sugar. The rate of photosynthesis is not only a quantitatively important process in that it determines how fast the plant can grow, it is also often a limiting factor in plant growth. An increase in carbon dioxide concentration in the atmosphere can thus lead to higher plant production. It acts as a fertilizer. In addition, there are secondary effects of increased carbon dioxide concentrations, such as more efficient water usage by the plants.

Different plants use different biochemical systems for capturing carbon dioxide and the response of a particular crop to the direct effects of an increased concentration in the atmosphere is determined by which group it belongs to. C_3-plants, to which wheat, rice and many other major crops belong, can take direct advantage of extra carbon dioxide, whereas maize, sorghum, cane sugar and other C_4-plants will benefit much less. There is a small group of plants with yet another photosynthetic system called CAM plants, pineapple being one example. These will not be affected by higher carbon dioxide levels (Figure 8.2).

In C_3-plants, carbon dioxide and oxygen compete with each other in two opposing processes—the trapping of carbon and the burning of it as fuel. An increase in carbon

Expected yield increase (ΔY) for a doubling of CO_2-concentration

C_3-plants
Wheat
Rice
Soybeans
Legumes
Trees
Weeds, etc

$\Delta Y = 10\text{-}50\%$

C_4-plants
Maize
Sorghum
Millets
Sugarcane
Prairie grasses
Weeds, etc.

$\Delta Y = 0\text{-}10\%$

CAM-plants
Sisal
Pineapple
Cactus, etc.

$\Delta Y \approx 0\%$

C_3 weeds in C_4 crops = problem

Fig 8.2. Direct effect of increased carbon dioxide on crop productivity

dioxide concentration changes the competition to the advantage of carbon dioxide trapping, which results in higher productivity and better yields at harvest. How big the effect is and if it will be sustained depend on if the plant has somewhere to store its energy, if there is the capacity to expand the number of seeds, for example. Experiments under artificially good environmental conditions have shown an increase of 36 percent in marketable yield in grain crops. In fiber crops, such as cotton, the yields have doubled. In C_4-plants and CAM plants, there is no competition between carbon dioxide and oxygen in the photosynthetic process, and the dynamics of the biochemistry do not indicate that those plants could take much extra advantage of higher carbon dioxide concentrations in the atmosphere

CARBON DIOXIDE FOR WATER CONSERVATION

In addition to the direct effect of carbon dioxide as a fertilizer for C_3-plants, increased atmospheric concentration may also shift the water balance of a plant. Crops exchange gases — breathe — through small pores in the leaves, so-called stomata. But as the stomata open, the plant also loses water to the atmosphere. In an atmosphere with increased concentrations of carbon dioxide, a plant will be able to capture the same amount of carbon dioxide with less stomatal opening or, if replacing water is not a problem, it will be able to capture more carbon dioxide than before. This advantage is shared by C_3-plants and C_4-plants, but it is difficult to quantify as it depends on field conditions such as wind and soil moisture. The estimated increase of yield in C_4-crops is 0–10 percent. For C_3-crops the combined effects would be an increase in growth and yield by 10 to 50 percent.

The higher water efficiency of plants in an increased carbon dioxide atmosphere could imply less risk of water stress. This may not be the case, however. It seems that the plants instead increase their leaf area, so that the water stress may still occur during dry periods but at a higher level of biomass.

A COMBINATION OF EFFECTS

The increased water efficiency in carbon dioxide-stimulated plants is an example of how carbon dioxide interacts with other environmental factors. It also seems that the carbon dioxide response is higher when there is lack of light, lack of enough nutrients or generally poor soils. This could make it possible to increase production in saline and polluted environments. However, the ability to enhance growth on soils with less nutrients, especially nitrogen, may also imply that the nutritive quality of the crops will be lower.

Temperature also enters the picture when calculating photosynthetic rates, even if it is difficult to quantify these effects and they may be counterbalanced by others. One effect is that the optimum temperature for photosynthesis increases by about 4–6°C, which would be an advantage in warm climates. The range of temperature optimum may also narrow, however and the growth rate would then vary more with environmental conditions.

We should also keep in mind that a scientist's glasshouse is not the same as nature and that increases in atmospheric carbon dioxide will also stimulate weeds that compete with the crops. Another complicating factor is that the speed of plant development may change in ways that make the time of flowering and seed setting less than optimal in relation to the climate. In some cases this could work to the advantage of higher yields, much depending on what crop varieties will be used in the future.

CLIMATE CHANGE AND A SHIFT IN RISKS

Studying the effects of climate changes on agriculture poses quite different problems compared to looking at the direct effect of carbon dioxide. Atmospheric carbon dioxide will increase at a steady and fairly predictable pace. Looking at climate it is usually not the average changes over a long time that are important in determining the success or failure of a crop. A 2°C-change in average temperature may not even mean much if the increase comes evenly. Farmers would be able to adjust by using new crops or other farming practices.

A shift-in-risk perspective might be much more relevant to the individual farmer. It deals with changes in frequency of disruptive droughts or unusually good years, for example. As climate changes, it is most likely at this level that the trends will be perceived. Two dry summers in a row. Spring being early and the first frost late. Again.

This perspective poses two problems when trying to formulate strategies for the future. One is the lack of information about the future climate on a regional scale. There are as yet no definite answers as to how frequent droughts will be or if heavy rains will become more common. There are only educated guesses that for some regions have a fairly good base in the climate models but for other regions lack consistent data to build the guesses on. The second problem is the difficulty in knowing whether a bad or a good year is part of a trend or only an unusual event. A particular year can therefore not form the basis from which the individual farmer can decide what should be grown the next year. And too many failures in a row will often be too much for the farm to handle economically. In subsistence farming, it may be the difference between having or not having anything to eat.

On the other hand, farmers are used to dealing with climatic risks as they are as much a part of today's reality as of tomorrow's. Preparing for and dealing with droughts in the future is not fundamentally different from strategies that can be adopted today. The difference and problem might be that climatic extremes become too common. Two hot summers in a row may mean real trouble even if there are strategies to cope with one. Several bad years in a row may lead to farmers having to abandon the land or giving up growing a particular crop. The replacement of traditional farming practices with higher-yielding more intensive agriculture in some developing countries may also have increased the vulnerability of a local system as some of the coping systems are no longer in place and have not been replaced by new ones.

What then is the increased risk of extreme climate in a warmer world? Based only on a shift in average precipitation, a calculation for England shows that a 1-in-a-100-year drought could become 7.5 times more frequent in any growing season with a doubling of carbon dioxide.

A FOCUS ON THE CROPS

The discussion of risk perception deals with the farmers' ability to adjust to changes in the environment. This is a factor that has not been taken into account in most of the studies of the sensitivity of the agricultural system that have been done so far. The focus has instead been on the impact on certain crops or on different regions. Summarizing some of the crop studies, two conclusions can be drawn about the change for the yields of wheat and maize in a warmer climate. In the crop regions in mid-latitude North America and Europe, warming with no change in precipitation will be detrimental to the yields of these crops. With a

2°C increase in temperature, average yields may fall by as much as 15 percent. A 10-percent decrease in North America would be equivalent to 10 percent of the global trade in cereals. Less precipitation would worsen the yields.

A closer look at the reasons for the decrease in yield puts focus on what factors will be important in determining the potential for agriculture in different areas in the future. The most important effect of higher temperature on plants is in increasing water loss. If there is not enough soil moisture to compensate for evapotranspiration, the plant will stop growing. Temperature also speeds up the growing process, which sometimes is an advantage but not always. The risk is that the development from seedling to seedbearing plant is accelerated in such a way that not as much of accumulated biomass is stored in the economically important seed. One study of wheat in England showed that this factor could reduce the yield by 7 percent.

In a colder climate, the effect of temperature will be different compared to places where low temperatures are not the important limiting factor for agriculture, especially if there is an abundance of soil moisture. Further north, the length of the growing season is limited by late spring frosts and early fall frosts. Here, a higher average temperature will give a longer growing season.

There will probably also be a shift poleward for some crops and, in mountainous regions, a shift upwards in altitude. One of the oft-quoted studies shows how the "corn belt" in North America will shift maize cultivation about 175 kilometers north to northeast for every 1°C rise in temperature (Figure 8.3). The altitude shift is illustrated in a historical study of how average temperatures have affected the risks associated with oat production in the highlands of Scotland. As temperature increased, some land that was only marginally suitable to grow oats on became economically feasible to use as the risk of crop failure

118

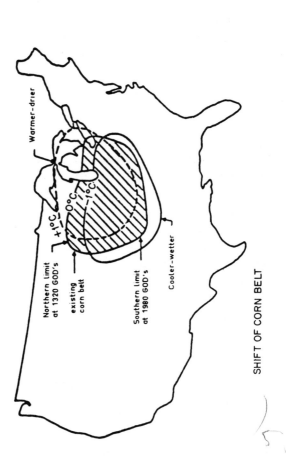

SHIFT OF CORN BELT

Fig 8.3. The length of the frost-free season is important for how far north maize can be cultivated. The figure shows the estimated impact on the US "corn belt" for a 1°C change in temperature

decreases. A Japanese study shows that rice production may well shift several hundred meters upslope in a climate compatible with a doubling of carbon dioxide. Summing up these trends, a 1°C increase in mean temperature would advance the thermal limit of cereal cropping 150–200 kilometers in the mid-latitude northern hemisphere and would raise the altitudinal limit by 150–200 meters.

SOILS AND PESTS

The simple exercise of moving climatic zones around on the map hides some limitations posed by other factors than just temperature and precipitation. In Canada, for example, poor soil quality may limit the poleward shift of agriculture. Another not-so-easily quantifiable factor is how increased rainfall may affect the storage of nutrients in the soil. Will the nutrients leach out at a faster rate leaving less for the crop? One study of the area around St Petersburg (formerly Leningrad) shows that soil fertility could fall by 20 percent with the increase in precipitation by the year 2035, in spite of increased use of fertilizer.

It is not only crops that will move polewards. Pests will surely follow the warming climate into areas where they have not been able to survive before. In some areas they will manage to produce two generations every season rather than one or manage to survive the winter better with higher population densities as a result (Figure 8.4). In the United States, the potato leaf hopper, which also affects soybean and other crops, will invade earlier in the growing season and thus cause greater damage. Soybeans in the grain belt may also suffer from more damage by earlier infestation of the corn ear worm.

Many livestock diseases are now limited to tropical countries but may spread in the future. Rift Valley fever

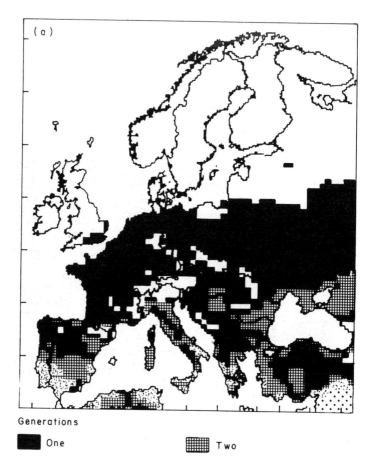

Generations

■ One   ▦ Two

Fig 8.4(a). Potential number of generations of the European corn borer at present-day climate

and African Swine fever are two examples. The horn fly may reduce weight gain in beef cattle and milk production in the United States, in addition to the losses of 730 million US dollars that it causes today. The Australian beef industry has previously had problems with ticks. Further stress on the

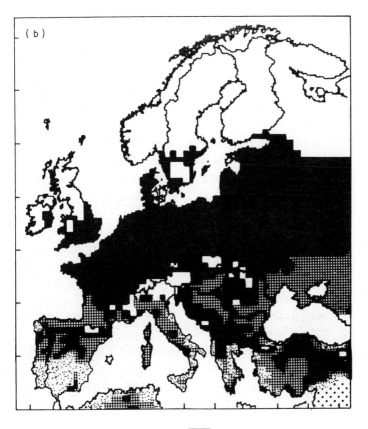

Three Three   Four Four

Fig 8.4(b). Potential number of generations of the European corn borer with a 1°C increase in mean annual temperature

cattle from long dry seasons in combination with climatic conditions that favor the ticks, may well increase these problems.

Migrating pests are quick to take advantage of extended climatic boundaries. Locusts thus reached new limits in Southern Europe in 1986–88. In New Zealand, increased locust swarming may well be an indication of climate change.

Cold regions, where pests at present pose small problems, may face completely new situations in the future. Iceland, for example, may come within the limits of the potato blight, from which it has so far been spared because of low summer temperatures.

It is not only temperature that affects the distribution of pests. Higher precipitation and humidity may favor fungal diseases that attack cereals. Aphids carrying viral diseases may also increase.

A SUMMARY BY REGION

How can current knowledge about the sensitivity of crops and livestock to climate be summarized? One answer would be "with difficulty" as regional climatic predictions are still unsure. In spite of this, the IPCC Impact Assessment has attempted a breakdown of the effects according to region, on which the following summary builds. The climate assumed is consistent with a doubling of carbon dioxide levels.

The Americas

In the United States, there will be a warming accompanied by a reduction in soil moisture, which will decrease the potential yields of maize. In California yields may drop up to 15 percent and on the Great Plains up to 25 percent assuming irrigation. Part of this may be compensated for by a small increase in yields in the Great Lakes region, but it will depend on how climate will affect soil moisture, which is one of the parameters that has been difficult to predict.

For wheat, the productive potential will also shift north and the direct positive effect of carbon dioxide on the plants will not help the farmers on the Great Plains enough to offset the lack of soil moisture. Dryland soybean yields will also decrease.

The ability to cope with the drier environment depends to a large extent on the possibility of offsetting the effects by irrigation. But in some of these areas the ground water table is already dropping and it may become more expensive to irrigate in the future as the replenishment of the aquifer will be even slower with reduced rainfall and snowfall.

Canada lies further north and may be able to reap some benefits from a warmer climate along the northern boundary of spring wheat cropping. However, the poor soil conditions and the terrain will limit the expansion of agriculture in areas other than in the Peace River District. In the main agricultural areas of Canada, there may be a decrease in yields because of the lack of soil moisture. The national figure is about minus 20 percent for the yield of spring wheat. Some of this may be replaced by winter wheat, which is better able to withstand spring and early summer droughts. For other crops such as maize, barley, soybeans, potatoes and hay, there may be a decrease in yields in most parts of Canada.

Mexico and Central America may also experience a decrease in soil moisture, which would decrease yields for rainfed crops such as maize. Irrigated crops may also be affected if problems with water scarcity become more common. Rainfall for this region is hard to predict, but an increase may mean more soil erosion, especially if the rains are heavy. The effect of rainfall may also be pronounced in Brazil where experience from the El Niño years of 1982–83 tells the story of reduced agricultural potential by almost 25 percent in the northeast part of the country. In other parts of Brazil, more rain could mean higher productivity for soybeans in the west and central parts and for wheat in the south. Some Brazilian crops, notably coffee and citrus fruits in the south, are occasionally destroyed by frosts. Such frost damage may be reduced in the future.

Looking at the patterns of precipitation in South America, one can expect an increase in areas that are moist today along the coast west of the Andes. The increase might be able to offset the effect of warmer climate on the soil moisture, keeping the land productive. The semiarid areas east of the Andes may become drier. The evapotranspiration in the Pampas region may increase by 1 percent, making it less favorable to raise cattle. Central and southern Argentina may gain instead, as the productive capacity of the grasslands can increase in a warmer climate.

The Andes themselves pose different climatic questions for agriculture. Here the restriction is the low winter temperatures at high altitudes. A 1°C warming may raise the cultivation limit by about 200 meters. More rain will favor potatoes at the expense of barley.

Europe

Some clear winners in a warmer world are the farmers of northern Europe. In Scandinavia, the yields of spring wheat could increase by 10–20 percent. The harvests of barley and oats would increase by similar amounts. In southern Scandinavia, some new vegetable crops may become feasible to grow which have previously been imported. On Iceland, agriculture is not a major activity, but sheep farming will gain from the increased carrying capacity of the grasslands.

In Europe in general, there will probably be a shift in agricultural potential from the south to the north. In the maritime areas of northwest Europe, including countries such as Ireland, the United Kingdom, northern France, the Netherlands, Belgium, and Denmark, more rain and higher growing-season temperatures will increase yields of grass and potatoes. Barley and maize may become profitable to grow several hundred kilometers further north than before as summer temperatures increase. Wheat does not benefit from the warmer climate, but the direct carbon dioxide effect may compensate for losses in yields due to high evapotranspiration.

Southern Europe will experience a decrease in productivity as temperatures increase and soil moisture decreases. Biomass potential in Italy is estimated to be 5 percent lower than today and in Greece 30–40 percent lower.

The European Alps have some similarities with the Andes in that temperature is determined by elevation. A change in climate projected with a doubling of carbon dioxide may be the equivalent of moving the Alps to the Pyrenees 300 kilometers south. Expressed in another way, the cultivation limit may move up about 500 meters.

Commonwealth of Independent States

Scientists in the former Soviet Union have used quite different methods to assess the impact of future climate on agriculture mostly relying on analogues with past climates. Their results indicate that with a 1°C warming (around the year 2000) there will be a decrease in productivity in the dry parts of the mid-latitudes, but the changes are not bigger than that they are within the limits of current agricultural adaptation to climate. They believe that further in the future, moisture conditions may improve and there is an increased potential especially for C_3-crops that can take advantage of the direct positive effects of carbon dioxide. A regional study shows that the agricultural potential may increase by as much as 10–20 percent with a warming of 2°C in all but two regions of the European part of the Commonwealth of Independent States.

The Middle East and Africa

The Middle East is a region where farmers are used to dealing with the limitations of a dry climate. It seems that conditions here will become even drier, but there are only a few studies that have looked at what impact it might have on the agriculture in the area. One Israeli study shows that wheat yields may be reduced by 40 percent.

The adjustments and limitations to farming in dry climates become even more apparent further south, which the recurring droughts in parts of Africa have shown during the past decades. Future problems can be illustrated by studies of Maghreb, where an increase in temperature by 1.5°C would increase evapotranspiration by 10 percent. This means considerably less water flow in rivers that are essential for irrigation. Some areas may no longer be usable for farming under future conditions, but instead may be converted to rangeland.

The climate in west and northeast Africa is dependent on the monsoon and a northward shift may increase precipitation in these areas. The net effect will depend on where it will rain more and whether it will rain enough to offset the increased evapotranspiration caused by higher temperatures. Any decrease in rainfall or a shorter rainy season would reduce yields of maize and the carrying capacity of rangelands. A look at the driest 10 percent of the years at present gives a reduction of maize yields by 30–70 percent and of forage yield by 15–60 percent. If the rains are intensified, erosion and flooding may become a problem in mountain regions such as in Ethiopia.

Eastern Asia

Flooding will probably become a significant problem in some already flood-prone regions of the world such as southern China and further south in eastern Asia. The summer monsoon will become stronger and move northwestward. Rice production could benefit from the increased summer rainfall. A Philippine study points to a 30-percent increase in yields with the combined effect of more rain, higher temperatures and the direct effect of carbon dioxide.

In some areas, the winter monsoon is as important for agriculture as it determines soil moisture during an important part of the growing season. The importance is illustrated by two estimates of the effect of climate change in China. With enough soil moisture, rice, maize and wheat yields may increase by 10 percent nationally, but if there is a lack of soil moisture there will be a decrease in maize yields by 3 percent for every 1°C warming. In south east Asia, reduced winter rainfall may decrease potential rice yields if the effect is not offset by the direct benefit of increases in carbon dioxide concentrations in the atmosphere. Too rapid growth and crop losses to pests would reduce yields of rice and maize.

The effects of climate change on Indian agriculture depend on what region one considers. In northern India, a 0.5°C higher temperature may mean 10 percent lower yields from the wheat harvest. In central India the situation is worse, unless rainfall also increases. On the positive side is the fertilizing effect of carbon dioxide on some plants. Sorghum, however, would not benefit as much from this and would also suffer from premature development of the seed, which would reduce yields.

For Japan, the positive picture is more clear as higher temperatures will enable the country to extend the areas that can be used for rice cultivation (Figure 8.5). Most of the

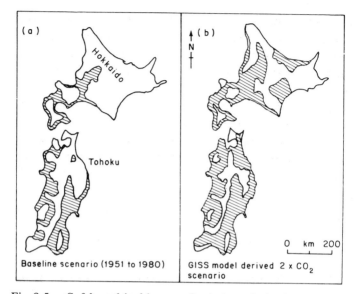

Fig 8.5. Safely cultivable area for irrigated rice in northern Japan under (a) current conditions and (b) with a climate expected with a doubling of carbon dioxide

gain will be made in the north-central region, but averaged over the country, the net increase in yields would be 2–5 percent. Yields of maize and soybeans will also increase, whereas the sugar-cane crops will suffer from reduced rainfall in the southern part of Japan.

The Pacific Islands, New Zealand, and Australia

In New Zealand, the effect of a warmer climate depends on what crops one studies. Wheat, barley and oats will lose while maize and vegetables may gain.

In Australia, the picture of the future of agriculture is far from clear cut. There are both gains and losses, much depending on changing patterns of rainfall. In general, decreases in crop production in one area seem to be balanced by increases in other areas and some crops will do better at the expense of others. The cattle industry may lose as the productivity of the grasslands decreases or when some grasses are replaced by less nutritious ones. Heat stress may also play a role, especially for sheep farming.

The small Pacific islands face quite a different situation from other countries in the area. Here the loss of land to the sea is the major problem for agriculture. Especially sensitive are crops grown close to the coast, such as copra. The potential yields of subsistence crops such as yams will also fall.

GLOBAL FOOD SECURITY

Looking at the rather drastic decreases in yields that have been implied for some of the bread baskets of the world, most notably the North American Great Plains, there has been a concern for international food security. Will there be enough food? What will happen to the surpluses that are used for emergency programs when the crops have failed in other parts of the world? What will happen to prices?

There is no compelling evidence that global food supplies will be radically diminished. The problem will be in the distribution of food production and in feeding a growing population. Some regions that are already vulnerable will probably suffer further food shortages. The areas pinpointed are the following — in Africa: Maghreb, West Africa, Horn of Africa, southern Africa; in Asia: western Arabia, southeast Asia; in the Americas: Mexico, Central America and parts of eastern Brazil (Figure 8.6).

Food security is not only a matter of available production, however. Much more critical for the distribution of food globally is its price. In the book *Climate Change and World Agriculture*, Martin Parry writes that a 10 percent decrease

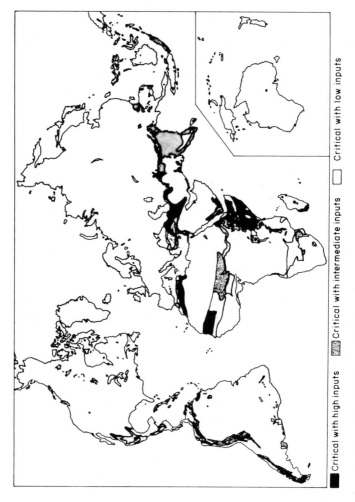

Fig 8.6. Regions identified as critical zones in respect to ability to support current populations

Critical with high inputs ■   Critical with intermediate inputs ▨   Critical with low inputs ☐

in yields may cause a 7 percent increase in prices, which can seriously limit the ability of food-deficient countries to pay for imports.

The basis for Parry's discussion is some of the few modelling studies that have been done to probe the impact of climate change on the global agricultural system. The models generally use estimated yield decreases for various crops and different regions and these numbers are then fed into computer programs that simulate the dynamics of the agricultural market as it looks today. Changes in farming practices are not taken into account, which makes it impossible to use the models for predictions. They are however useful in looking at the sensitivity of the market to changes in yields.

One example is the International Future Simulation Model run by the U.S. National Center for Atmospheric Research (Figure 8.7). Looking at the year 2000 and feeding in a 20-percent decrease in yields, the total agricultural productivity only changed 5–7 percent. The interpretation is that the world agricultural system, according to this model, has the capacity to absorb about two-thirds of the change in potential by adjusting the intensity of agriculture, land area under production, what crops to grow etc. It would, however, require a slow and steady change.

More concern is raised by model runs where abrupt changes in yields were entered. With a 20-percent decrease in yields in one year, the crop reserves were reduced to almost zero. This was followed by an overcompensation in production the next year, where the model showed a glutting of the market and a collapse of prices. An environmental situation causing this scenario would be droughts in all the major grain-exporting areas in the same year. Parry concludes that changes in the frequency of droughts and warm spells may become critical to global food security.

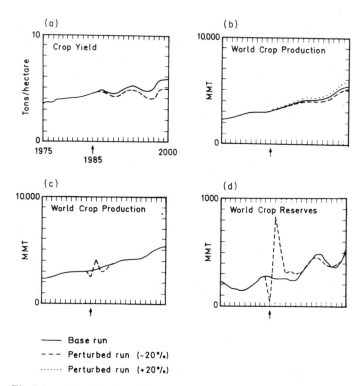

Fig 8.7. Simulated agricultural effect of a perturbed climate. (a) Crop yield in the United States with a slow decrease in yield by 20 percent. (b) World crop production with slow change. (c) and (d) Effect on world crop production and reserves with a sudden single decrease in yields

Another model uses regional changes in food production as input together with three different future climate scenarios. This USDA Static World Policy Simulation does not, as the name implies, take dynamic responses such as feedback loops, into account and is thus not very realistic for looking at a market. Nevertheless, it gives a clue to the sensitivity of a system.

In a climate situation consistent with a doubling of carbon

dioxide, there was little change in prices and food production capability. Favored regions in Australia, China and the former USSR compensated for losses in North America and Europe. If, on the other hand, several major grain-producing regions were hit with decreasing harvests, the impact on prices could be pronounced. In a worst case scenario, with a 10-percent reduction in the US, Canada and the EEC, a 25-percent reduction in developing countries and no compensating increases in yields, prices would go up 20 percent. If the US, Canada and the EEC were even harder hit, with a 50-percent yield reduction, prices would go up 50 percent.

It should be emphasized that these studies do not take into account the farmers' ability to take advantage of new technology and new farming practices, which with a changing climate may well be promoted.

REGIONAL ECONOMY

Aside from looking at the direct effect on prices, a few regional studies have focused on the effect of a changing agricultural sector on the economy of a region. The method here is to link models of biophysical, economic and social impact and look at a cascade of events. For example, model estimates for agricultural yields in a region under certain climatic conditions are fed into a farm simulation model. Here, it interacts with agricultural practice, such as fertilizer use, and gives an output in production and prices. These, in turn, are fed into an economic model that looks at how the economy of an individual farm might affect other industries and employment in the area.

Taking all the unknowns and all the model limitations, the results cannot be regarded as a prediction of what will happen in an area with a certain climate scenario, but the approach places a focus on the fact that farming is an integrated part of the economy in a region.

One regional study has dealt with the Province of Saskatchewan in Canada. Farming in the northern part of this region is currently limited by the number of frost-free days. With a doubling of carbon dioxide, the climate here will change. In the climate model used, there is an increase by 3.4°C in growing-season temperature and an increase of precipitation by 18 percent. This would mean a 50 percent increase in the number of days when temperatures are high enough for cultivation. A predicted increase in precipitation could have meant more available soil moisture, but that effect is offset by the higher temperatures. The number of droughts even increased.

Adding up the numbers, the potential for biomass production increases. In the southern part of the province the increase ranges from 1 to 30 percent and in the northern part it is more than 70 percent. So far so good. The problem is that the major crop in this area, spring wheat, will not be able to benefit from the improved climate. Rather, the models predict a decrease in production by 25 percent. It is the southern part of the province that is hit, an area that today contributes 13–14 percent to the world wheat trade.

The smaller harvests hit the economy of the farmers and total farm income decreases by the same amount as the yields in the models. The result looking at the regional economy is a decrease in farm employment by 3 percent. The total employment in the province decreased 2 percent and the total provincial gross domestic product fell 12 percent. If the farmers are not able to make any adjustment, this would imply that a decrease in yields could have a rather large effect on the economy of the region.

ADJUSTING TO CHANGE

One of the important lessons from the experiments with models of agricultural yields is probably that adjustments will be necessary to counterbalance any negative impact of a changing climate. What adjustments are possible? What can farmers do? How should governments use agricultural policy?

Along the northern limits of cultivation, new land can probably be taken into use for intensive agriculture. This will be the case in northern Europe and the northern part of Russia. In Canada, poor soil may limit this possibility. Some high-latitude protected valleys in Alaska, northern Japan, southern Argentina and New Zealand may also become feasible to cultivate as well as some high altitude regions. A limit to altitude expansion may be that some of this land, especially in the Alps, is currently used as pasture.

New crops can probably be used. High-yielding varieties that have temperature limitations for where they can be grown, may expand. An example is that yields of quick-maturing rice might be able to increase by about 4 percent if the number of growing days increase in northern Japan as is predicted in scenarios for a doubling of carbon dioxide-levels. Using late-maturing rice in southern Japan might increase yields by 26 percent.

Drought-tolerant crops will certainly be in demand in the future as many otherwise fertile regions will suffer greater risks of drought. Irrigation will also become more important than today. The question for the individual farmer will be whether it is more economical to irrigate at increasing costs or to shift to more drought-tolerant crops. Irrigation efficiency will become very important as water resources in many areas are scarce and the competition for the water will become even greater than today.

Irrigation systems are often big investments and cannot change as easily as some other farming practices. It is easier to adjust when and how the soil is plowed, sowed and harvested, in the amount of fertilizer used and on when it is applied. The possibilities of fine-tuning the system for optimal use in a future climate will to a large extent depend on how fast changes occur and how well they can be predicted from year to year.

Despite the possibilities to adjust agriculture to a changing climate, the future will hold some great problems. A major cause for concern is that salinization, desertification, and general soil erosion will diminish the amount of land available for food production, regardless of any climate change. In addition, the land has to feed more and more people.

CUTTING EMISSIONS

Coping with climate change is not the only reason farmers need to adjust their practices. Equally important is to cut agriculture's share of the increase of greenhouse gases. On a global scale, this share is substantial. Methane, nitrous oxide, and carbon dioxide from agriculture together contribute 14 percent of the greenhouse gases in the atmosphere today. The main sources are flooded rice cultivation, the use of nitrogen fertilizer, improper soil managament, land conversion and biomass burning.

Without necessary measures, emissions from agriculture will probably increase in the future as population grows, more land is cleared and the area used in paddy rice cultivation increases. Additional rice cultivation itself may contribute with a 35-percent increase in methane emissions. In addition, the use of nitrogen fertilizer is increasing and nitrous oxide emission from this source may increase by 70–110 percent by the year 2025.

This trend of increasing greenhouse-gas emissions is not inevitable, however, and changing agricultural practices will be important in reducing emissions. One focus has been on rice cultivation, where reduction of methane emissions by 10–30 percent is possible by paying attention to the management of water systems, cultivar development and efficient use of fertilizers. To reach this goal, it is necessary to gain a better understanding of the dynamics in the rice field and the relationships between methane production and other factors. The theoretical reduction also does not take into account any increases in cultivated area.

In trying to reduce methane emissions from livestock production, scientists already have some of the basic understanding of the processes involved and they know that environmental goals can go hand in hand with increasing productivity. The quantitative effects depend on how intensively managed the animals are, but the emissions may be reduced 25–75 percent per unit meat or milk.

One measure that can be implemented in the short term is supplemental feeding with locally produced additives, which may become important for livestock on poor diets. In India, systems for supplemental feeding have been tried and shown to be a self-sustaining and economic investment. A more controversial issue is the use of growth-stimulating drugs, such as bovine somatotropin, to increase the efficiency in feed conversion.

In the future, more research on diet modification of intensively managed animals could have great impact on methane emission. Whole cotton seed and polyunsaturated fats have, for example, been shown to affect the metabolic process in a favorable way. More knowledge is needed about what farming systems would be able to take advantage of these kinds of processes. Another future possibility is to change the microbial flora in the intestines of the animals. There are some studies that show how a reduction of the protozoa in the digestive system of ruminants lowers methane emissions and enhances productivity, but more knowledge is needed to judge if this approach would be practically and economically feasible.

AGRICULTURE AND DEFORESTATION

In the tropical regions of the world, biomass burning and deforestation are major contributors to the increasing concentrations of greenhouse gases. Measures to reduce the need for further deforestation are thus important. They could include enhancing the role of indigenous uses of the native forests and to establish new forest cropping systems. Agroforestry has the advantage of enhancing storage of carbon in both the soil and in the canopy. It also stabilizes the soil and provides firewood. This approach has multiple benefits aside from the environmental ones, which is a fact that will be very important to the success of any attempt to reduce emissions of greenhouse gases.

Similar benefits can be the result of increasing the productivity of the existing croplands by efficient use of water, fertilizer and appropriate crop varieties. This would also reduce pressure to break new land. Using fertilizers with slower conversion rates or nitrogen-fixing crops are ways to cut nitrous oxide emissions, for example. More efficient agricultural practices may also reduce the need to burn crop residues to add nutrients to the soil. In a similar way, the routine burning of the savannah can be replaced by using improved forage species and management systems. Restoring degraded agricultural land is another possibility that can be pursued.

The chance for all of these measures depends to a large extent on policy decisions and on opportunities to learn about new options in farming. There is a need to evaluate land-tenure policies to provide incentives to limit conversion of forest to agricultural systems and to start education programs to teach about the consequences of soil degradation.

CUTTING EXCESS NITROGEN

In temperate climates, the major agricultural problem is the increasing and often excessive use of nitrogen fertilizer. In addition to adding nitrous oxide to the atmosphere, leaching of the excess fertilizer in some areas pollutes groundwater as well as coastal areas. Reasons to control fertilizer use better are therefore two-fold and the technology to do it is available. It is more a matter of awareness and to some extent regulatory practices. One suggestion that has been made is to develop "balance sheets" based on best available knowledge to decide allowable limits for nitrogen use.

CARBON IN SOIL

Another discussion of response strategies involving agricultural practices deals with carbon flux: where and under what circumstances carbon is stored in or released from the soil. Whether the field is plowed, for example, influences how carbon is stored or released. Using crop rotation with organic amendments can increase soil equilibrium levels of carbon. According to a World Bank working paper, the theoretical potential of extra carbon storage in the soil is immense, of the same magnitude as the net increase of carbon dioxide, but much more research is needed to establish the dynamics of the system in relationship to other factors, such as nitrogen fertilizing, and it is too early to make concrete recommendations.

Chapter 9

Forestry

ALMOST A THIRD OF EARTH'S LAND AREA IS covered by forests. From an aerial view of the world, they can be seen as two green belts. One of them encircles the equator, the other takes up large parts of the temperate and subpolar regions in the northern hemisphere. The forests are important in two ways in the context of climate warming. Firstly, they store carbon, thus keeping carbon dioxide away from the atmosphere. Destruction of forests is emptying those stores and contributing about 10 percent of the emissions of greenhouse gases. Secondly, a warmer climate will affect the potential for forest growth and regeneration. This in turn may affect the ecology of the forests as well as the economy of regions that are dependent on forestry. From both perspectives — forestry as a villain and forestry as a victim — there is a need to develop strategies for coping with new demands.

MAPPING VEGETATION TYPES

One of the classical approaches to looking at the effects of climate change on vegetation has been to move vegetation zones around on a map based on temperature and moisture indexes. A dry and cold region would be covered by tundra vegetation, for example and a wet, warm region by tropical rain forest. This kind of life-zone classification, originally developed by Holdridge after whom it has been named, has been refined and applied to a climate warming perspective. It is easy to envision how the deciduous forest typical of continental Europe and large parts of the United States would move north to replace coniferous trees. The boreal forest in turn would move further north taking over part of the tundra. In the south, the deciduous forest would have to fight an expanding landscape of dry bushy vegetation. The gain would probably not compensate for the loss, however.

This simplified version of just moving vegetation types around on a climate map lacks an important dimension: time. It does not take into account the dynamics of forest growth. It only looks at the potential for different vegetation types but neglects adjustment time. It thus neglects a fundamental difference between forestry and most agriculture. The generation time of most agricultural crops is one year and changes can be made from one year to another. The trees in a forest may live for several hundred years and even a highly managed forest can have turnover rates as high as 50–100 years, which poses limitations on the ability to adapt to changes. But changes will occur in both managed and natural forests and there are models that attempt to simulate the dynamic response of forests to look at what future the forestry sector of society will have to adjust to.

THE TIME OF DEATH AND RENEWAL

How a forest grows is determined by the increase in size of individual trees and by new space becoming available for seedlings to develop. This process of tree death or seedling development is very unlike the smooth continuous tree growth, but they can both be incorporated into a model that gives a picture of how a forest will respond to climate change (Figure 9.1). This approach has been used to model forest responses since the most recent glaciation in a variety of forests on different continents. One example is a simulation of the change in forest composition in the Cumberland Mountain Rim in Tennessee during the past 16,000 years. The model results have been validated by comparing them to the pollen record from Anderson Pond, Tennessee, U.S.A.

One important result of model studies of climate change is that there is a time lag in the response of the forest.

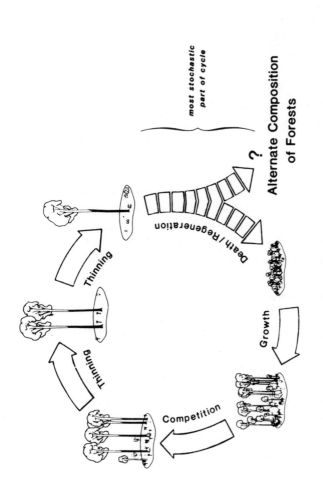

Fig 9.1. Forest regeneration and growth as depicted in one model of forest productivity

Only when an old tree has died, is there a chance for a new tree to develop and take advantage of new climatic conditions. Based on this fact a simulation of northern hardwood forests in the US showed that it would take a hundred years before the forest came in phase with the climate. This means that in a continuously changing climate, the forest would grow under less than optimal conditions over large areas.

There have also been some more extensive experiments conducted with this kind of model trying to look at how the forest will spread over a continent in response to climate change. Looking at eastern North America with a doubling and quadrupling of carbon dioxide, the result was a slower growth of deciduous species throughout much of their geographical range. Many dominant species disappeared completely, especially in the transition between boreal and deciduous forest. In a warmer climate one might expect the deciduous forest to move north, but the presence of boreal species delayed that response in the model experiment.

The concept of lag time is partly the result of the rate at which various species can migrate. Those rates might determine whether the maps of vegetation types are relevant for a relatively rapid climate change. Will a species keep up with the change in climate and be able to stay within temperature ranges to which it has adapted? Or, will it be hopelessly left behind and subject to climate stress making it more vulnerable to pests and disease?

One attempt at answering these questions has been made by looking at fossil pollen records of the warming periods after the last ice age. The migration rate then was generally 25 to 40 kilometers per century as seeds spread northward with wind, water and animals. As a comparison, in mid-continental North America each degree of temperature change corresponds to a distance of 100–125 kilometers.

This could mean a 300 to 375 kilometer northward shift of the temperature zones in the next century. It will thus be very difficult for the natural forest to keep up with the pace of climate change. A study of the distribution of hemlock in northeastern US and southeastern Canada illustrates this problem (Figure 9.2). It shows that the actual area covered by this tree species will decrease rather drastically in a future warmer climate as it is not able to advance northward at the same pace as it will disappear from its current southern limits.

Species migration will of course be affected by humans. In managed forests, the selection of seedlings may speed up the change. In areas where the forest has to compete with other land uses and migration routes may be cut off, the changes may be delayed.

PESTS, POLLUTION AND SOIL QUALITY

There are some factors other than climate directly that will have large effects on forest dynamics and productivity. One is the distribution of pests, another the effect of air pollutants, both of which could hinder the establishment of new forest or speed up the dieback of old forest. Examples of pests that may increase are the spruce budworm in Eastern Canada and bark beetles in northwestern American forests. Some such changes have already been attributed to climate change. In Australia, higher temperatures and more precipitation may make *Pinus radiata* plantations more sensitive to disease. Extreme weather events, such as destructive storms, may also become more frequent in some areas and affect forest productivity.

Fire is also important. In a warmer world, the amount of biomass would be higher making more fuel available if a forest fire should start. In addition, many forested areas

Fig 9.2. Present and future range of eastern hemlock under two climate scenarios. Light diagonal shading is the present range, the dark diagonal shading the potential change with CO_2 doubling and the cross hatched area of overlap is where the trees are likely to be found 100 years from now. (a) Predicted by Hansen et al., 1983.

will become drier making it more likely that a fire can start. In Canada, the area hit by forest fires doubled during the 80s compared to the previous decade, and there have been speculations that a drier climate has contributed to this. Even if it is too early to make connections to climate change it shows how drier sites can give fuel to large forest fires.

The ability of forest vegetation to respond to climate change also depends on factors that have nothing to do with

Fig 9.2.   cont. (b) Predicted by Manabe and Wetherald, 1987.

climate. One such factor is soil quality, for example the soil's ability to hold moisture. A model simulation of a site in northeastern Minnesota is a case in point (Figure 9.3). With soil of high water-holding capacity, spruce trees were replaced by a more productive northern hardwood forest. These trees could take ample advantage of the warmer climate. In addition, the forest litter was easily decomposed, which increased nutrient availability amplifying the temperature-induced increase in productivity. In total, there was a 50 percent increase in above-ground carbon storage. Running the same simulation with a sandy soil, which does not hold water well, gives a very different result. Here the

150

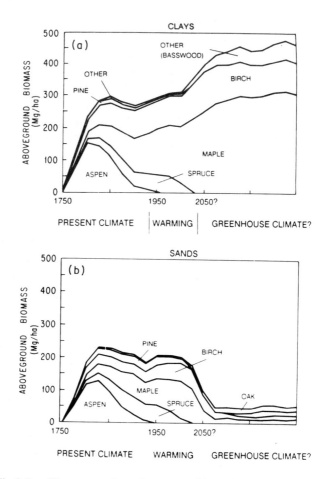

Fig 9.3. Biomass and species composition of Minnesota, USA, forest. The climate in both graphs is as predicted with a doubling of carbon dioxide. (a) shows the simulated growth on soils with high water-holding capacity and (b) the results on soil with low water-holding capacity

spruce forest is replaced by stunted pine-oak forest with only 25 percent of the carbon storage compared to the original forest.

It should be evident that the composition of forests we see in the world today is the result of many different factors and the present distribution of natural ecosystems is a function of a unique combination of climate and other ecosystem characteristics. Fossil records show that plant community types may well disappear over time as new ones take over. The old-growth Douglas fir forest is, for example, only 6,000 years old and represents only 5-10 generations of the longer-lived tree species.

A HIGHER OUTPUT?

Fire, pests and reduced soil quality could all contribute to lower output from the forests that are economically important for timber or paper production. But is there no potential for higher output in a warmer world? Would a longer growing season not be an advantage? In some cases, the answer is probably yes, but it depends on if temperature is the limiting factor for forest productivity. In many areas, the growth is limited by the lack of nutrients in the soil and unless one can compensate by large-scale fertilizing of the forest, higher temperatures would not increase productivity.

The direct impact of carbon dioxide might also be interesting to look at. As most other plants, trees are stimulated by higher carbon dioxide concentrations in the atmosphere because there is more carbon available to bind in the photosynthetic process. In addition, there is less risk of losing water when trying to capture the carbon dioxide through the stomata or ''breathing pores'' in the leaves or needles. The theoretical understanding of carbon dioxide stimulation is good, but it has been very difficult to quantify the effect.

Most experiments on carbon dioxide stimulation have been conducted with single leaves, shoots or small seedlings and none of the studies have looked at what happens in the field under the long-term influence of higher carbon dioxide levels. Looking at how the stomata of different species of trees respond to changes in carbon dioxide concentration, one can guess that the largest short-term increases in carbon assimilation will be in conifers. Lack of water or other environmental stresses may offset this positive direct effect of carbon dioxide, especially if mid-latitudes become less humid, as is predicted in several of the climate-change scenarios. In general, the higher carbon dioxide concentration will increase water-use efficiency, however.

More important than any short-term response to carbon dioxide is whether the actual growth rate of trees will increase in a way that is sustained year after year. Even small differences in growth rate are important as they accumulate over several years and result in large differences in final plant size. The studies that have been done suggest that the growth rate may indeed increase, but the magnitude of the change varies with different species and is also dependent on other environmental factors.

Where does the extra carbon go? Most of it seems to go to the roots and shoots under ground, but there may also be more branches, tillers, flowers and fruits on a plant. In addition, there may be changes in the chemical composition of the plant tissue, such as higher carbohydrate levels. This, in turn may influence the interplay between the plant and predators and pests, but it is too early to speculate how.

Moving up in complexity to considerations of the whole forest, it becomes even more difficult to quantify the effects of higher carbon dioxide levels. One would expect a higher rate of biomass production as long as it is a young forest. As the ecosystem ages, there will eventually be a balance of carbon dioxide uptake by the growing vegetation and carbon dioxide release by plant respiration and decomposition of dying matter. However, a lack of quantitative data on carbon dioxide response at this level in combination with lack of data on how the trees in a forest influence each other, for example in creating microclimates, make it impossible to build any reliable theory of forest production in a high carbon dioxide world.

One of the major critics of climate warming, Sherwood Idso, has studied carbon accumulation in sour orange trees over several years and claims that the increase in carbon accumulation in the trees is sustained. Therefore, he suggests that a potential increase in carbon storage in the world's forests should be considered in the mathematical models of the carbon cycle. Other scientists are, however, wary of his conclusion that carbon dioxide stimulation would be important enough to offset the projected increases in emissions of carbon dioxide from burning of fossil fuel and deforestation.

SUMMARIZING IMPACTS

Despite uncertainties in the present approaches of looking at climate change and forestry, one can conclude that a doubling of carbon dioxide can cause considerable changes in the composition, areal extent and location of the forests of the world. The largest changes will probably occur in the temperate and boreal regions. The tropical forests will probably be more sensitive to changes in precipitation than to temperature, but there is not enough climate data to make any suggestions as to the directions of the change.

Going one step further in the descriptions of future problems in forestry, there are some specific changes to expect when physical stress on the forest increases. One example is from the US Great Lakes region, where climate-induced forest decline will start to become apparent

30–60 years from now. Within the same time span, changes will also be seen in the dry areas of the central and western US and in 60–70 years southeastern US forests will be affected. In Australia, the climate habitat suitable for *Pinus radiata* and *Eucalyptus regnans* may completely disappear.

PROBLEM OR POSSIBILITY?

The forests of the world constitute a great carbon reservoir. The amount of carbon stored is at present equivalent to the amount in the atmosphere. Historically about one-third of all forests have been lost as humans have broken the land for agriculture and other activities (Figure 9.4). Today the forest clearing takes place mostly in the tropics (Figure 9.5). The fact that forest burning and felling contribute 15–30 percent of the emissions of carbon dioxide to the atmosphere puts focus on the need for policy decisions that could turn forests into a sink of carbon rather than a source. How can the destruction of tropical forest be limited? Can replanting of new forest become a quantitatively important carbon trap that could actually help stop the increase of greenhouse gases?

The issue of tropical forest clearing is highly political and sensitive as it is often posed as a conflict between global environmental interests and the day-to-day survival of people relying on forest clearing for creating fields for food production. There are, however, policy options that can serve both needs. The most important one is replacing forest clearing with sustainable cropping systems for land that has already been cleared. It is a matter of choosing crops, fertilizing, and working the soil appropriately as discussed in the chapter on agriculture. Long-term research in the Peruvian Amazon indicates that improved agricultural practices that result in a sustainable agriculture may save

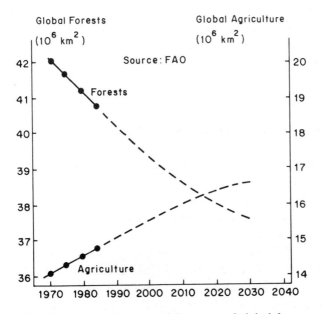

Fig 9.4. Variation with time of the areas of global forest and global agriculture assuming no significant retardation of the ongoing deforestation

a forest area that is five to ten times as big as the improved field.

Tropical forests are not only cleared for agricultural land, however. The timber industry is economically important for several tropical countries and also contributes to the greenhouse effect because of current destructive harvesting techniques. The most important remedy is to develop harvesting practices that are not as damaging as the ones used today. Currently 30–70 trees are often left damaged while 2–10 trees are removed per hectare.

Planting new forest could be an option for providing local fire wood and in the context of agroforestry. For the timber industry this is only a long-term option, however, as it takes

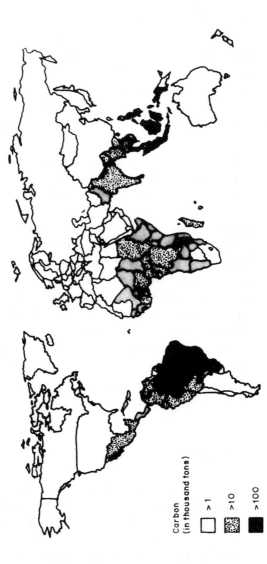

Carbon
(in thousand tons)

☐ >1

▦ >10

■ >100

Fig 9.5. Carbon dioxide emissions from tropical deforestation in 1989

30–70 years before the wood would be of the same quality as that of the natural forests and therefore harvestable.

Forest use in the temperate and boreal regions does not today pose the same conflicts as in the tropics, even if the temperate regions historically are responsible for most of the forest loss. Generally, northern forests are now considered net carbon sinks rather than sources. Pollution-induced forest death might shift that balance and contribute to the greenhouse effect. The policy issues deal with measures to maintain productivity in the face of further forest stress in a changing climate. Cutting down on air pollution would be an important measure. In managed forest it could also include the choice of species to plant, for example.

In some cases it might be possible to increase productivity, making the forests into more efficient carbon sinks. A study of Finnish forests concludes that in the coming 50 years, 270 million tons more carbon could be stored in the above-ground biomass by using appropriate forest management. This amount is equal to Finnish carbon dioxide emissions in the coming 40 years. If a temperature increase is added to the scenario, the gain in productivity equals the total use of fossil fuels in Finland during the next 60-70 years. This best-case scenario shows that there are possibilities even if all of them may not be realized.

REFORESTATION

Planting new forest has been suggested as a way to catch carbon dioxide in the atmosphere to balance anthropogenic carbon dioxide emissions. To reach this goal one would have to plant an area equivalent to the size of Europe from the Atlantic to the Urals. Replanting on a smaller scale is more realistic and such measures could reduce greenhouse

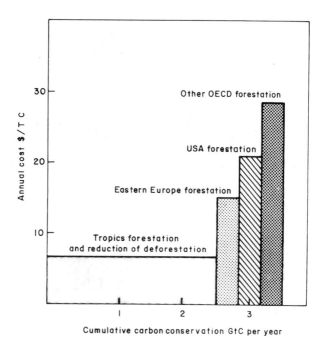

Fig 9.6. Regional average cost of forest management measures

gas emissions by 10–15 percent. In Australia, a plan is already underway in which one billion trees are to be planted. The major purpose of the project is soil conservation but the possibility of capturing carbon dioxide is an added plus.

The cost of large-scale tree planting is difficult to estimate as it varies depending on local environmental and economic factors. A rough global estimate ranges from 200 to 2,000 U.S. dollars per hectare (Figure 9.6). There are also legal problems of ownership of the land to be reforested.

Chapter 10

Natural ecosystems

FFECTS OF CLIMATE CHANGE ON AGRICULTURE and managed forests can be considered special cases of how a warmer world will change the natural life on Earth. Most ecosystems are not as highly managed, however, and their economic worth not as apparent as the incomes derived from crops, for example. The species and vegetation types that make up natural ecosystems will nevertheless face the same kinds of changes as have been discussed in the chapters on agriculture and forestry.

To get a feel for the magnitude of the changes, we can compare the greenhouse-gas-induced warming to earlier changes in global climate and see that a doubling of carbon dioxide will give a global climate that is warmer than at any time within the past 200,000 years. During this time, there have been drastic changes in the geographical patterns of terrestrial ecosystems due to smaller fluctuations in climate. For example, during a warm period from AD 800 to 1200, when temperatures around the North Atlantic were only 1°C warmer than today, the Canadian boreal forest was found much further north. On Iceland and in Norway, cereal cultivation flourished. During another warm period, in the early Holocene 9,000–6,000 BC, current subtropical dry zones were much wetter than today and there was productive savannah grassland where there is desert now. During the earliest warm peak before the last ice age, around 120,000–80,000 years ago, temperatures were 2–3°C higher than today and large areas now occupied by boreal forest were covered by deciduous species.

These scenarios cannot be used directly to make projections for the future, but they tell the story of the ability of vegetation to adapt to climate change and how different species can take advantage of new opportunities. A major difference today compared to past climate changes is the rate of change, which will occur 15–40 times faster than previous natural fluctuations in climate. In the short-term perspective, the fast pace might well become important in determining how ecosystems will or will not adapt and move in response to the changes.

WHO WILL COPE AND WHO WILL NOT?

Maintaining species diversity in all parts of the world has become an important environmental priority in international discussions. The focus has been on the destruction of habitats, for example in the tropical rain forest. In the long run, climate change may also become an important threat to the survival of some species.

Who will be able to adapt or move in response to the changes? Who will not be able to cope and instead disappear? One important factor is the dispersal capacity of the species, for example the distance seeds can travel with wind, water or animals. Poor dispersers are much more likely to become extinct if the climate change is fast, while opportunistic weeds may quickly take advantage of new life room opening up. In an ecosystem, different species may have different capacities for dispersal and thus move at different paces in response to the climate. The result would be new mixes of species in the ecosystems that will develop in a warmer climate.

Small or isolated populations of plants or animals will also have difficulties in adjusting. Species that are specific to a certain island will, for example, not always be able to respond by moving to a more suitable climate. Isolated populations can also be genetically impoverished with little evolutionary potential to adapt to the changing environment. Species that reproduce slowly have a similar disadvantage. Based on these considerations, one can expect an increased risk for extinction of species on islands, on mountain peaks and in remnant vegetation patches in an otherwise developed landscape. Peninsulas, such as the Yucatan Peninsula, can almost be considered a biogeographical trap where wetter conditions may eliminate areas of dry woodland and bush.

Natural reserves and heritage sites are also risk areas. From an ecological point of view they are isolated in a manner similar to islands as they are surrounded by a developed landscape, to which the protected species cannot migrate. In general, natural corridors of migration are often cut off by agricultural land, roads and human settlements, which will hamper migration of species to suitable climate zones.

A similar risk for species decline can be expected in climate zones that will diminish on the whole or in some cases almost disappear. A case in point is the Siberian tundra. The boreal forest may well advance northward so that the tundra will to a large extent be replaced by forest, with a loss of many tundra-specific species. Tundra nesting habitats for migratory shore birds may also be affected by arctic warming.

Most vegetation studies focus on the shift poleward of climate zones. There may also be a shift inland for some species that are today limited to coastal maritime climate zones. Examples are frost-sensitive species such as holly and bell heather that today have eastern boundaries in Europe that are correlated with cold winter temperatures. They will probably move eastward. Mild winters can also be detrimental. Many tree species in the boreal forest, and tall grass herbs are dependent on a certain number of frost days. To survive they will have to migrate eastward into western Siberia.

There seems to be much less research work done on wildlife populations compared to vegetation types. Most animals are probably more mobile than the plant communities, but an easy shift cannot be taken for granted. For example, migratory birds that would be considered extremely mobile are sometimes dependent on specific resting sites along the migration paths. If those sites disappear, the bird population might well be affected. Animals adapted to a polar climate face a general decrease in living space. Some of them will be able to adapt by moving further north, but increasing competition for food and other resources may diminish some populations. A special case is the polar bear, which is dependent on sea ice for travelling and feeding. A reduction in sea ice may place the polar bear population at risk.

Not all species will suffer from changes in climate. Some may gain, especially species that are specialized in colonizing new areas where old vegetation has been removed by fire or drought. In some cases the invading species are problematic, as they can spread in an uncontrollable manner. An example is a bamboo-like Australian plant, which already forms dense stands in the Florida Everglades, where the natural ecosystem has been destroyed by drought and fire. Another potential weed are symbiotic nitrogen-fixing shrubs that can colonize bare ground and arrest further development of the plant community for some decades. In wetlands, species that are temperature-limited may expand their geographical ranges. Water hyacinth, water lettuce, purple loosestrife and African pyle can thus be expected to become problematic weeds further north.

ONE STRESS ADDED TO ANOTHER

Vegetation and animal populations will not only be exposed to changes in temperature and precipitation. Equally important will be how these in turn affect the water balance and hydrology of a region. Wetland habitats may be especially susceptible to such changes and there are studies of the Florida Everglades that show how subtle alterations in the hydrology have profound effects on the population stability of wading birds. This in turn can influence energy flow and general ecosystem functions. Another example is how sphagnum species in bogs can be sensitive to changes in the hydrology. They have a key role in the accumulation of vegetation matter in the bogs.

In semi-arid climates in the Mediterranean, new precipitation patterns may cause salination of the soil. The problem is connected to the lengthening of the dry season and an accompanying increase in irrigation. In extreme cases of salination, the land is no longer suitable for agriculture and turns into desert.

Summarizing the effects of climate change on natural ecosystems, it is important to remember that this environmental stress is added to other problems caused by human activities. In industrialized areas, air and water pollution are probably of more concern when trying to protect biological diversity. In heavily populated areas or areas with expanding human populations, the major threat is the competition for land, where too little is often left for some plants and animals to survive on in the long run. The problem is that a rapid change in climate can aggravate these already existing problems, sometimes even making conservation efforts futile. An environment that is stressed by inappropriate climate in relation to the species that make up the ecosystem, will also be more vulnerable to other environmental stresses.

Chapter 11

Water, snow and ice

CLOUDS FILLED WITH WATER. A DOWNPOUR. A dam filled to the brim, almost overflowing. A dry and empty riverbed. A blanket of white snow over the landscape. Ancient ice frozen in the ground.

Water in its many forms provides a basic sustenance for life on Earth. From the chapter on agriculture, its role for vegetation and human food production should be apparent. The water in Earth's hydrological cycle also plays a crucial physical role in the environment and for human activities. Streams and rivers transport silt from inland to the coast. Heavy rains can cause mudslides. Lack of rain changes the characteristics of the soil. Rivers and lakes form major transportation arteries. Snow provides protection from cold air. Snowmelt fills the water reservoirs used for hydroelectric power. The frozen ground in permafrost regions keeps biological processes in check and stores vast amounts of greenhouse gases.

Will these roles change in a warmer world? How does climate affect the hydrological cycle? What critical processes determine if water will stay frozen or turn into liquid form?

Some of these questions are easier to pose than to answer. Water runoff has to be studied on the regional or even local level. The scale is the size of the river basin and climate models do not give reliable results. It is thus difficult to make any statements as to how the flow in a certain river basin can be affected by climate change. There are, however, methods to study the sensitivity of the hydrological systems that can highlight some cause-and-effect relationships between precipitation patterns and runoff.

SENSITIVE RIVERSHEDS

One conclusion that has been drawn about runoff sensitivity is that river watersheds are very sensitive even to small changes in climate. This is especially true in arid and semi-arid regions where the rainfall is unevenly distributed over the seasons and can vary greatly from year to year. A case in point is northern Africa, including the Sahel, where precipitation has been more than 50 percent below average since the 1970s. This may be part of a natural climatological cycle, but the effects are nevertheless severe. The reduction in runoff can be more than twice the reduction in rainfall. The effect can also be seen in Lake Chad, which since the 1960s has shrunk more than 11 times in size. The decrease in precipitation not only makes water scarce, it also reduces water quality and increases the risk of pollution problems as wastes are not washed away.

Situations less extreme but similarly sensitive to changes in precipitation can be found in parts of North America and southern Europe. If the weather becomes warmer and drier, the water resources in many regions of the US are expected to decrease by 1.5 to 2 times. In dry regions, such as in the southwestern US, changes in runoff could be amplified five times as compared to the changes in precipitation. In some regions, including the United States, some of the quantitative effects on water resources can be moderated with the extensive reservoir systems that exist but lower flow will still affect water quality.

In humid areas, the response in runoff is not as marked. Nevertheless, areas that will experience increases in precipitation can also expect higher flood peaks. The effect on people and environment will depend on if reservoirs will even the flow. In some cases flooding can be aggravated by forest clearing as the vegetation further up the river basin will no longer have the same capacity to capture the water.

It is not always precipitation that is the critical factor. In watersheds where snowmelt is an important source of runoff, changes in temperature may have great impact. In the mid latitudes of the northern hemisphere, a moderate warming of 1–2°C will increase winter runoff drastically and spring runoff will decrease as the snow will have melted earlier in the season. An example of the impact of such changes can be found in a Norwegian study. The conclusion is that the intensity of the spring flood will decrease substantially while winter runoff will increase as the snow cover will be present for one month instead of three. This could lead to erosion problems unless there are changes in agricultural practices, as the water carries away soil that is bare in the winter. On the positive side are the effects for hydroelectric power, where the water flow will be more evenly distributed and could help increase power output by 2–3 percent.

In large hydrological systems, different interests often compete for the water resources. Great rivers are an example where changes in precipitation will affect how much water for irrigation can be taken out without depleting the water source for people living further downstream. Another case is large lakes where maintaining the water level is important for navigation while shoreline developments are vulnerable to too high water levels (Figure 11.1) . The Great Lakes of North America are one such system that has been extensively studied (Figure 11.2). An expected drop of water level will have large economic impacts for the industries and municipalities around the lake as hydroelectric power generation will decrease and commerical navigation will have to make costly adjustments.

WHO WILL HAVE A WHITE CHRISTMAS?

One change in precipitation and temperature that can have large consequences is if snow will instead fall as rain. It is very difficult to make any predictions about seasonal snow

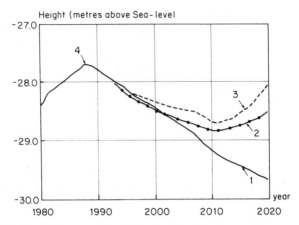

Fig 11.1. Estimated level of the Caspian Sea under varying climate change scenarios. 1 — stationary climate with man's impact. 2 — model-based anthropogenic change in climate including man's impact. 3 — map-based anthropogenic change including man's impact. 4 — observed lake-level variations before 1988

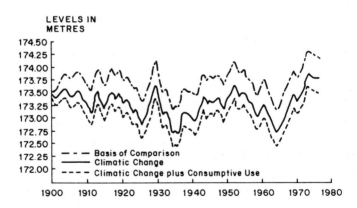

Fig 11.2. Comparisons of actual level of Lake Erie for the period 1900–80 with estimated levels for the same period generated by superimposing onto the actual conditions a warmer climate and a warmer climate with future level of consumptive use

cover in a warmer climate as there are several feedback mechanisms involving changes in albedo that are important in this context. Moreover, very small differences in temperature can cause a transition from snow to rain or vice versa. In general, the areas covered by snow will decrease in both the northern and southern hemispheres. In some areas the depth of the snow cover may increase, however, especially over Antarctica and over Greenland.

Snow has an important ecological role as it insulates the ground from the cold air. With less snow, some crops and natural perennials will thus risk frost damage in areas where they have previously been protected. The snow that accumulates in the winter also provides water resources for agriculture and hydroelectric power and changes in the availability of water are to be expected. For agriculture, earlier summer drying of the ground can become an important factor in increasing the risk of water stress and drought.

Snow cover is also of great economic importance for the tourist industry. Many communities in the Alps could be severely hit if the ski slopes could no longer operate or if the winter season becomes much shorter than today. Calculations for southern Quebec in Canada show that the number of snow days suitable for skiing could be cut by half. The economic loss would be tens of millions of dollars. In the south Georgian Bay in Ontario, the 50 million-dollar tourist industry may disappear completely.

Local as well as global climate is also affected by the snow cover. The snow reflects incoming solar radiation keeping it from heating the air as much as would be the case with a less reflective surface. Less snow implies a greater warming of the air than would otherwise be the case.

ON FROZEN GROUNDS

If snow cover is mostly a seasonal phenomenon, the opposite is true for permafrost. Somewhere between a fifth and a quarter of the land surface contains permanently frozen water that has been locked up since the last ice age. There is also permafrost beneath the seabed in the Arctic Ocean.

The permafrost will not melt immediately in response to above-freezing temperatures, but since much of it exists close to its melting point, it will react to changes in climate before reaching a new equilibrium. The result will be a deeper unfrozen and biologically active layer of soil over the permafrost. If the warming is great enough the permaforst will eventually disappear, even if this can take centuries. There are, however, thaw-sensitive areas that can react much faster. An historical example is the Mackenzie Valley between the 1800s and 1940s, where a 3°C rise in temperature moved the permafrost boundary several hundred kilometers.

Changes in the permafrost can cause instabilities in the soil as can be seen in pot holes and much larger structural disturbances that are associated with its thawing and freezing. This has to be taken into account when constructing roads, buildings and dams in these areas and when assessing the integrity of existing constructions. The natural ecosystem will also change. Landslides can cause uprooting of vegetation locally. Large areas with a deeper biologically active layer will become available for colonization by deeper rooted trees. The forest will thus change character with more broad-leaved species and denser coniferous stands.

A thawing permafrost may also provide a positive feedback loop in a warming climate as more ground becomes available for biological activity. If the ground remains waterlogged, microbial methane production will increase and can add significantly to the greenhouse gases in the atmosphere. In addition, there is methane trapped in water beneath the permafrost. If this is released as the permafrost degrades, it could raise the global temperature by another 0.4°C by 2020 and by 0.6–0.7°C by the middle of the next century.

How much of the permafrost will melt due to climate warming? The figures are uncertain but calculations for the former USSR show that a 1°C warming may cause the southern boundary of permafrost to move north and northeast by 200–300 kilometers. A 2°C warming could displace the boundary by 500–700 kilometers at the same time as the depth of the biologically active layer will increase where the permafrost remains. Similar trends can be projected for Canada. Mexico has permafrost areas on some high mountain peaks. With a 2°C warming those could disappear, with consequences for the local hydrology and ecology.

Chapter 12

Answering the challenge

IT IS GETTING COLD OUTSIDE AND THE HOUSE IS chilly. Should I turn up the heat? Is it worth the money to invest in insulating the house? I live quite a way from my work place. Should I take the bus and commuter train, or will I drive? My answers to these and similar questions are of no trivial importance. Together with other people's efforts to cut energy consumption, it is the core of any change toward cutting emissions of carbon dioxide and other greenhouse gases. Other everyday decisions can be added to the list. Do I carry bottles and cans to recycling centers? Do I plan my work so as to not travel unnecessarily? Am I willing to support tax policies that discourage use of fossil fuels even if it cuts into my income?

Individual decisions are not enough to change the present increase in emissions, however. If I take the commuter train depends on if it exists at all. The energy efficiency of taking the bus depends on if the engine is tuned and on what fuel it runs on. Does it burn fossil-fuel-derived diesel or maybe ethanol made from biomass? Those decisions might be made by the local bus company and in turn determined by fuel taxes set by politicians. My willingness to recycle depends on if someone can take care of my old bottles, cans and newspapers, if local politicians have prioritized it or if it is a profitable business to pursue for someone.

INTERNATIONAL POLITICS

Politics on the national level will not be enough to deal with the increases in greenhouse gases. Emissions from one country will always affect the global atmosphere and, no doubt, the negative impacts of climate warming will hit countries that have not contributed to the increase in greenhouse gases while some heavy polluters might not suffer that much. From the start, climate warming has therefore been an international issue and it is widely recognized that international agreements on limiting emissions are necessary. The question is what form and content those agreements should have.

There are several political issues that are the focus of attention in trying to negotiate a global climate convention. The most important is how responsibility should be shared between industrialized and developing countries. Historically, and at present, the industrialized world is responsible for most emissions, particularly emissions from burning of fossil fuels. But the scene is changing. Deforestation in the tropics is also an important contributor to the greenhouse effect. In addition, the use of fossil fuels is increasing in many developing countries and will skyrocket with population growth, increased industrialization, and economic development unless new sustainable energy systems can be used. One immediate question is thus who should pay to make new, energy efficient technology available for developing countries. How much are the industrialized countries willing to put in an international fund for technology transfer?

Some industrialized countries may not even be willing to pay for their own changes. The United States, in particular, has taken a position that it will be cheaper to adapt to a changing climate rather than restructuring the whole energy system. This may be true for a specific region, but hardly on a global scale and there is a great risk that the most severe effects of climate warming will hit people who can least afford to adapt. This will eventually also hit the more fortunate regions as it is likely to cause migrations of environmental refugees.

In discussing the international politics of reducing greenhouse gas emissions, the question of how to divide the responsibility has quickly become important. Carbon dioxide emission can for example be calculated per capita, dividing the atmosphere equally among the citizens of the world. But from what point in time should one count? If previous emissions are to be entered into the calculations, industrialized countries clearly would have to cut their emissions drastically to allow for some growth in the developing world.

The next question is how to divide Earth's capacity to absorb carbon dioxide. The oceans and some forests, notably in the temperate region, are effective sinks for carbon dioxide. Should they be considered a common global resource or should each country be allowed to subtract the absorbing capacity of their territory from their emissions? One might say that the industrialized world has already used more than its share of the common sinks and should therefore have to pay for any further use. The concept of polluter pays could form the base for a system of tradable emission permits, which would be one way of transferring money for new technology to developing countries.

Regardless of what system the world community can agree upon in sharing our common atmosphere, it will be crucial to come to some agreement about cutting emissions of greenhouse gases. Once such agreements exist, we can focus our attention on how to conserve energy and actually reduce the emissions.

ENERGY POTENTIALS TO TAP

The technical potential to cut emissions of greenhouse gases is significant, especially in the energy sector. According to IPCC, this sector accounts for 57 percent of the radiative forcing from anthropogenic sources today. By using existing cost-effective technology, global emissions could be cut by 20 percent by the year 2020. The strategies would include improving energy efficiency and conservation. But even if these measures are cost effective, realizing their potential would require intervention by government policies.

A 20-percent reduction would not be enough to stabilize the concentration of greenhouse gases in the atmosphere, however. That would require additional energy conservation as well as fuel substitution and the marginal cost of every new reduction could be high. The specific cost for a country or industrial sector will depend on development of new technology, global energy markets and the time it takes before politicians start implementing control policies.

There are examples of the potentials within the energy, transportation and industrial sectors that show that quite a bit can be done. Most buildings are not as efficiently insulated as they could be. Energy requirements could be cut by half in new buildings and by 25 percent in older residential houses. In commercial buildings, the technical potential to cut energy cost is even higher — 75 percent in new buildings and 50 percent in old ones. The barrier to this potential being realized is economic. Energy-efficient technologies will need help from government policies and other institutions to be able to compete on the market. Improved maintenance of buildings would probably be economical, however and would require education more than anything else.

Cars, buses and trucks have become much more fuel efficient since the world was struck by the oil crisis in the 1970s and there is still technical potential for improvements. The rate at which more efficient cars can be counted on also depends on the rate at which old cars are substituted for new ones. Improved driver behavior with softer driving at lower speeds and vehicle maintenance can be encouraged immediately, however, as well as promotion of public transport.

Another possibility is fuel substitution. Ethanol derived from renewable biomass does not contribute to the greenhouse effect as the carbon dioxide released is part of a closed system where the next crops will catch it. The main problem is that traditional fossil fuels are too cheap for new alternatives to be able to compete.

In the industrial sector, substantial energy savings are possible as exemplified by the dramatically increased energy efficiency due to better processes and better design and materials in some industries over the past 15 years. The IPCC estimates the technical potential to be anywhere from 13 to 40 percent depending on industrial sector. The rate at which these changes can be implemented internationally depends to a large extent on economic factors. The extremely run-down and energy-inefficient industry in Eastern Europe and the Commonwealth of Independent States is a case in point.

Fuel switching also has some potential in industry, for example burning biomass wastes instead of fossil fuels or using methane recovered from landfills. Another option, which is increasingly pursued, is recycling of materials that require energy to produce. Some countries recycle the majority of the aluminum cans that are sold. Glass recycling and the use of recycled paper are other options, the latter being important in conserving forest resources.

Generation of electrical power can also be made more efficient. There is a potential for 15–20 percent better efficiency, but the rate at which that can be realized depends on what rate old power stations are replaced by new ones. Shifting from fossil fuels to other energy sources can also be important, but if hydroelectric or nuclear power is to be used as substitutes, other factors such as health risks and ecological consequences have to be taken into account. Substituting oil with natural gas could cut carbon dioxide emission by 30 percent.

The list of energy-saving measures could easily be made longer as exemplified in the IPCC report of Response

Table 12.1 Examples of short-term options (From *IPCC Response Strategies*)

I. IMPROVE EFFICIENCY IN THE PRODUCTION, CONVERSION, AND USE OF ENERGY

Electricity generation	Industry sector	Transport sector	Building sector
• Improved efficiency in electricity generation: —repowering of existing facilities with high efficiency systems; —introduction of integrated gasification combined cycle systems; —introduction of pressurized fluidized bed combustion with combined cycle power systems; —improvement of boiler efficiency • Improved system for cogeneration of electricity and steam • Improved operation	• Promotion of further efficiency improvements in production process • Materials recycling (particularly energy-intensive materials) • Substitution with lower energy-intensive materials • Substitution with lower energy-intensive materials • Improved electromechanical drives and motors • Thermal process optimization, including energy cascading and cogeneartion • Improved operation	• Improved fuel efficiency of road vehicles: —electronic engine management and transmission control systems; —advanced vehicle design: reduced size and weight, with use of lightweight composite materials and structural ceramics; improved aerodynamics, combustion chamber components, better lubricants and tire design, etc.; —regular vehicle maintenance;	• Improved heating and cooling equipment and systems: —improvement of energy efficiency of air conditioning; —promotion of introduction of area heating and cooling, including use of heat pumps; —improved burner efficiency; —use of heat pumps in buildings; —use of advanced electronic energy; management control systems • Improved space conditioning efficiency

•Introduction of photovoltaics, especially for local electricity generation
•Introduction of fuel cells

trucks;
—improved efficiency in transport facilities;
—regenerating units
•Technology development in public transportation:
—intra-city modal shift (e.g. car to bus or subway);
—advanced train control system to increase traffic density on urban rail lines;
—high-speed inter-city trains;
—better intermodal integration
•Improved driver behaviour, traffic management, and vehicle maintenance

—improved heat efficiency through highly efficient insulating materials;
—better building design (orientation, window, building, envelope, etc.);
—improved air-to-air heat exchangers
•Improved lighting efficiency
•Improved appliance efficiency
•Improved operation and maintenance
•Improved efficiency of cook stoves (in developing countries)

continued overleaf

Table 12.1 (*continued*)

II. NON-FOSSIL AND LOW EMISSION ENERGY SOURCES

Electricity generation	Other sectors
•Construction of small-scale and large-scale hydro projects	•Substitution of natural gas and biomass for heating oil and coal
•Expansion of conventional nuclear power plants	•Solar heating
•Construction of gas-fired power plants	•Technologies for producing and utilizing alternative fuels:
•Standardized design of nuclear power plants to improve economics and safety	—improved storage and combustion systems for natural gas;
•Development of geothermal energy projects	—introduction of flexible-fuel and alcohol-fuel vehicles
•Introduction of wind turbines	
•Expansion of sustainable biomass combustion	
•Replacement of scrubbers and other energy-consuming control technology with more energy efficient emission control	

III. REMOVAL, CIRCULATION OR FIXATION

Energy/Industry	Landfills
•Recovery and use of leaked or released CH_4 from fossil fuel storage, coal mining	•Recycle and incineration of waste materials to reduce CH_4 emissions
•Improved maintenance of oil and natural gas and oil production and distribution systems to reduce CH_4 leakage	•Use or flaring of CH_4 emissions
•Improved emission control of CO, SO_X, NO_X, and VOCs to protect sinks of greenhouse gases	•Improved maintenance of landfill to decrease CH_4 emissions

Strategies (Tables 12.1 and 12.2). Many of these technologies have been evaluated and proven to be functional. Much bigger question marks are if they are economically feasible and what market potential they might have. The answer will depend on the national context and on the industrial sector but policy decisions at the national and international level will also be important in how technical options are evaluated. Will protective patents hinder the diffusion of technology or could the system be made to help it? What international programs could help poor countries to get access to efficient but more expensive technology? How would one best promote education about energy conservation, nationally and internationally?

THE PROCESS HAS STARTED

The task to reach any international consensus on reduction of emissions of greenhouse gases may seem immense and very difficult to tackle. This is especially true as it requires cooperation from many actors in all parts of the world and in all sectors of society. There is no simple technological fix to the problems we face. But the process has started and some important steps have already been taken. One is the international consensus to cut emissions of CFCs as spelled out in the Montreal Protocol on Substances that Deplete the Ozone Layer.

The Montreal Protocol was originally signed by 24 countries in 1987 as a response to mounting scientific evidence on the destructive potential of chlorinated and brominated halocarbons to the stratospheric ozone layer. The appearance of the ozone "hole" over the Antarctic and a general thinning of the ozone layer had made scientists and policy makers acutely aware that human-kind has the ability to alter the global environment.

Table 12.2 Examples of medium-/long-term options (From *IPCC Response Strategies*)

I. IMPROVE EFFICIENCY IN THE PRODUCTION, CONVERSION, AND THE USE OF ENERGY

Electricity generation	Industry sector	Transport sector	Building sector
• Advanced technologies for storage of intermittent energy • Advanced batteries • Compressed air energy storage • Superconducting energy storage	• Increased use of less energy-intensive materials • Advanced process technologies • Use of biological phenomena in processes • Localized process energy conversion • Use of fuel cells for cogeneration	• Improved fuel efficiency of road vehicles • Improvements in aircraft and ship design —advanced propulsion concepts; —ultra-high-bypass aircraft engines; —contra-rotating ship propulsion	• Improved energy storage systems: —use of information technology to anticipate and satisfy energy needs; —use of hydrogen to store energy for use in buildings • Improved building systems: —new building materials for better insulation at reduced cost; —windows that adjust opacity to maximize solar gain • New food storage systems that

eliminate
refrigeration
requirements

II. NON-FOSSIL AND LOW EMISSION ENERGY SOURCES

- Nuclear power plants:
  - passive safety features to improve reliability and acceptability
- Solar power technologies:
  - solar thermal;
  - solar photovoltaic (especially for local electricity generation)

Advanced fuel cell technologies

- Other technologies for producing and utilizing alternative fuels:
  - improved storage and combustion systems for hydrogen
  - control of gases boiled off from cryogenic fuels;
  - improvements in performance of metal hydrides;
  - high-yield processes to convert ligno-cellulosic biomass into alcohol fuels;
  - introduction of electric and hybrid vehicles;
  - reduced re-charging time for advanced batteries

III. REMOVAL, RECIRCULATION OR FIXATION

- Improved combustion conditions to reduce N_2O emissions
- Treatment of exhaust gas to reduce N_2O emissions
- CO_2 separation and geological and marine disposal

Furthermore, it has brought attention to the risks of increases in ultraviolet radiation for human health and food production.

The focus was first placed on the ozone destructive potential of the CFCs and other stable halocarbons but their effect as greenhouse gases have added to the concerns. Since 1987, the Montreal Protocol has been considerably strengthened and now it calls for a complete phase-out of the worst CFCs. Industrialized countries would be required to achieve this by the year 2000, and developing countries by 2010. It also cautions about increases in some compounds that are not as destructive but that can become problematic, such as HCFCs and methyl chloroform. A further strengthening is possible the next time the protocol is to be revised, in November 1992. The protocol has gathered quite a following on the international scene. More than 90 countries have signed it including many developing countries. For some countries it is still controversial, however. China only recently signed the protocol and India is still hesitant.

On the technical level, CFCs in many uses are already being replaced by new compounds and new processes that are more environmentally sound. One example is the replacement of CFCs as a propellant in spray cans, another as a cleaning agent in the electronics industry. The agreement to phase out the destructive chemicals has also spurred research and technological innovations and it will probably be possible to meet some of the very highly set goals. In some cases there are costs involved, in others there have actually been savings.

The process by which the decision to phase out CFCs has been made is important for the discussion of climate change. It shows that the world can gather around a shared environmental goal. Not all problems of technology transfer to developing nations and economic help to effect the change have been solved, but the process is under way. In addition, there are efforts within the United Nations Environmental Programme to continuously evaluate new technical options that can facilitate the phase-out.

Climate change is a bigger issue. The science is more complex and involves not just one class of compounds but many. The sources of greenhouse gases are not as easy to quantify as for CFCs, which are purely man-made. Moreover, the effects in terms of a warmer world, have not shown themselves. Yet. The problem is that action cannot wait. Just as with CFCs, the problem grows as we wait. Any policy measures to decrease the effects have to be more drastic the longer we wait to implement them.

SCENARIOS OF CHANGE

What, then, are the options? What effects would different policy decisions have? One type of answer to these questions is provided in the IPCC Response Strategies, where the authors have constructed different scenarios ranging from business-as-usual to one with accelerated control policies (Figures 12.1 and 12.2). A closer look at those scenarios show both the possibilities and some of the obstacles that have to be overcome. They also put different sources of greenhouse gases — energy, agriculture and forestry — into perspective.

The major result is that if emissions continue to grow with no changes, the concentration of greenhouse gases expressed as carbon dioxide equivalents would double by 2025. With a low-emission scenario, the carbon dioxide doubling would be reached 15 years later, 2040. To stabilize concentrations at a level that is below a doubling of carbon dioxide would require accelerated control policies.

There are some common assumptions to all the scenarios. One is that global population levels will rise according to

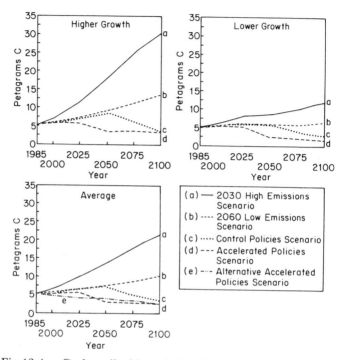

Fig 12.1.   Carbon dioxide emission from fossil fuels assuming different levels of economic growth. The time for doubling of carbon dioxide has been revised to 2025 instead of 2030 and 2040 instead of 2060 since the publication of the IPCC report

estimates made by the World Bank. This increase will result in a global population of 9.5 billion by 2050 and 10.4 billion by 2100. More people by necessity means more land used for agriculture and food production. The IPCC thus assumes an increase in all the variables that are associated with agriculture, mainly methane emissions from rice cultivation and cattle farming and nitrous oxide emissions from increased use of nitrogen fertilizer. The two more stringent scenarios, "control policies" and "accelerated control policies", require a change in fertilizer use, new rice cultivation practices and methane inhibiting techniques in the meat and dairy industries.

An increase in agriculture and deforestation might go hand in hand if there are no efforts to stop current forest clearing for food production by making more efficient use of already cleared land. Equally important is to change destructive logging practices. The scenarios that eventually result in a stabilization of carbon dioxide concentration require a slowdown of deforestation and reforestation of previously cleared land.

Turning to industry's potential, the most immediate contribution will be made by decisions to phase out CFCs. To stabilize concentrations of greenhouse gases in the atmosphere requires that all countries that have signed the Montreal Protocol phase out CFCs completely by the year 2000 and that there are limits placed on emissions of carbon tetrachloride and methyl chloroform.

ENERGY DEMAND AND ECONOMIC GROWTH

The most important contribution to moderating any climate change has to be made in the energy sector. Energy demand is closely linked to economic growth, which makes an increased energy demand seem a necessity if the developing world is to reach a living standard comparable to the industrialized countries. The IPCC scenarios calculate using two different assumptions. In their high-growth scenario, some regions in the developing nations reach an economic level that is typical of Western Europe in the 1970s by 2020. In the low-growth scenario, this development is delayed until 2075. Other regions remain below this level of development in both growth assumptions. In the business-as-usual scenario, it makes a huge difference if economic growth is high or low. If, on the other hand,

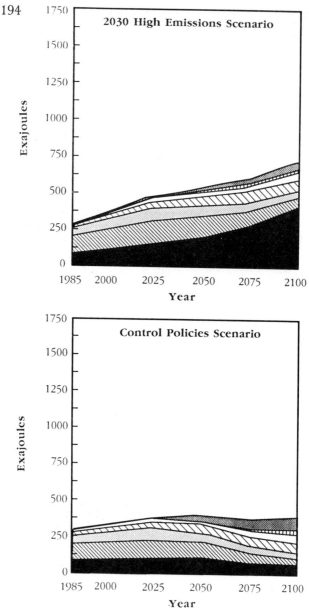

Fig 12.2(a).   Global energy supply at lower growth

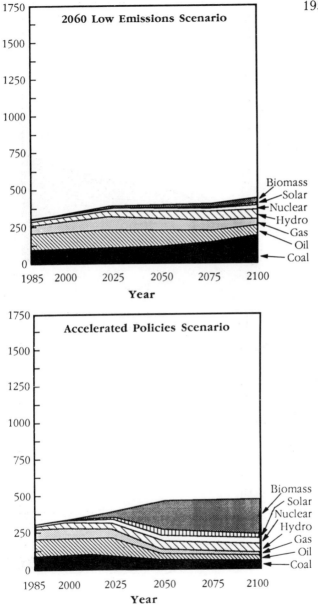

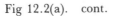

Fig 12.2(a). cont.

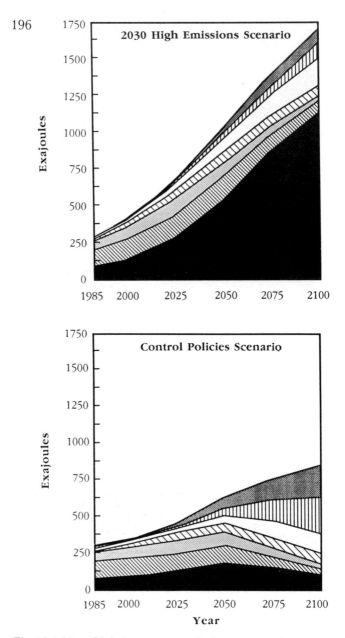

Fig 12.2(b). Global energy supply by type at higher growth

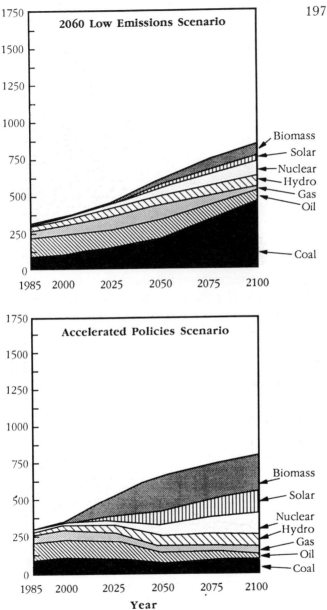

Fig 12.2(b).   cont.

control policies concerning energy use can be implemented, energy consumption only has to be about half of what would otherwise be the case with the same economic growth. As the remaining growth in energy demand can be met by renewable resources, a growth in global economy does not necessarily have to lead to higher carbon dioxide emissions. It will, however, require political decisions about control policies to reach such a goal. Some countries have specific plans to reduce the emissions but these countries are responsible for only a minor portion of the total worldwide emissions (Table 12.3).

The most important decision is to improve energy efficiency in existing systems. That would not be enough, however. Non-fossil fuels also have to start playing a larger role. To reach the goals set in the accelerated control policies scenario, solar electricity and biomass energy have to be commercially viable early in the next century and have to have a significant impact by 2025. In addition, costs and other constraints for nuclear power are assumed to diminish. Another assumption is that there are fees based on the carbon content of fossil fuels (50 US dollars/ton coal) implemented on a global scale by the year 2025.

In spite of the close coupling between economic growth and energy use, there are also some energy advantages with better economic development. For example, industrial machinery and other energy-consuming technologies can be more quickly replaced by more energy-efficient versions. Economic growth could also give new opportunities for incomes in some countries that today destroy forests by overharvesting.

WAIT AND SEE?

Many policy makers might agree on stringent energy conservation measures if they were convinced that global

Table 12.3   National carbon dioxide targets

West Germany	25% cut by 2005
Netherlands	Stabilize by 1994 and 3–5% cut by 2000. Then substantial cuts thereafter
Denmark	20% by 2000 and up to 50% by 2030
Italy	Stabilization by 2000 and possible 20% cut by 2005
France, Belgium, Luxembourg	Support a Community proposal to stabilize emissions at 1990 levels by 2000
UK	Stabilize at current levels by 2005. Wide range of measures in Patten Report (1990)
USA	Bills pending in Congress for 20% cut (Oregon already has legislation for 20% cut by 2005)
Switzerland, Norway, Sweden	Freeze at current levels. Sweden plans CO_2 tax (1991)
New Zealand	20% by 2005
Canada	Stabilize CO_2 emissions at current levels by 2000
Japan	Stabilize at 'lowest possible level' by 2000
Second World Climate Conference —European Community, Australia, Austria, Canada, Finland, Iceland, Norway, Sweden, Switzerland	Stabilize at 1990 levels

Source: WWF

warming and climate change are realities to count on in the near future. They would start acting if the costs were proven to be too high to justify inaction. But the scientists are not 100 percent sure, someone might say. The regional effects cannot be predicted yet. There are also scientists who do not believe that the models are worth trusting or that feedback mechanisms can counterbalance any increases in the greenhouse effect. And if climate warming is a reality, it might not be that costly to adjust. Why not wait until we have all the answers? Would it not be better to put our efforts into other problems?

The problem is that the cure will be tougher the longer we wait. The authors of the IPCC-report provide an illustrative example: In all scenarios, carbon dioxide emissions have to decline in order to achieve stable concentrations at the doubling of carbon dioxide by 2090. If we start in the year 2000, a gradual reduction of 17 percent is sufficient. If measures are not implemented until 2020, however, the model calculates that a reduction of 60 percent (from 2020 levels) is necessary to achieve the targets.

Another, complementary answer is that reducing emissions of greenhouse gases is not in conflict with other environmental goals. CFCs have to be phased out to prevent climate warming but also to hinder further destruction of the ozone layer. Improving the efficiency of combustion processes for fossil fuels not only reduces carbon dioxide, but also reduces emissions of sulphur dioxide and nitrogen oxides, which play an important role in acidification. Switching energy sources can in some cases further reduce emissions. Cutting excess use of nitrogen fertilizer in agriculture would limit emission of nitrous oxide, but also prevent further eutrofication of sensitive coastal waters and nitrogen leaching into the groundwater. Recycling can save energy but it also saves natural resources and it can keep our waste dumps from overflowing.

A closer look at any environmental problem today provides incentives to start being conservative with resources and energy and to start being radical in creating environmentally sound policies.

Chapter 13

A game of science and politics

My main responsibility is not to be too impressed by the extremes in either direction and to continue with the systematic evaluation of new information and of what we know today.

THIS STATEMENT COMES FROM THE CHAIRMAN of IPCC's scientific committee, spelling out his philosophy in leading the tricky interplay of science and politics in preparing for an international climate convention. His name is Bert Bolin and he is a professor of meteorology at the University of Stockholm in Sweden.

Ever since 1973 when Manabe presented the first models of climate change due to increased carbon dioxide in the atmosphere, Bert Bolin has been convinced that climate warming will be a reality that has to be dealt with. But he gets apprehensive when people make statements that are not based on science, ". . . like when people talk about a five-meter rise in sea level, which is a lie, and when someone makes statements about crop failures in countries where we know nothing about such risks, for example in the United States. There is a risk, but we can't paint the picture as if we knew."

What we know is bad enough, according to Bert Bolin, and he has spent most of his scientific career either studying the dynamics of the carbon cycle and the increases in carbon dioxide or convincing international organizations that the problems caused by fossil-fuel burning have to be taken seriously. Already in 1958, he made his first prediction of carbon dioxide levels at the turn of the century. His estimates were a bit low but not too far off from what we will actually see. At that time, carbon dioxide levels in the atmosphere were only a research question, but the scene changed at a climate conference in Stockholm in 1974, when Manabe presented his first model of climate change.

"It was a remarkable result that to a large extent is still valid today. There have, of course, been much more detailed calculations but the conclusion is basically the same.''

A few years later, in 1979, it became clear that the climate-moderating capacity of the sea would probably delay the possibility of actually recording global warming and the need to make a concerted effort to critically evaluate the problems became acute. Bert Bolin had an opportunity to talk to the head of United Nations Environment Programme, Mustafa Tolba and a ball started rolling that resulted in a report on the greenhouse effect (SCOPE 29) and a meeting in Villach, Austria in 1985.

''Tolba's introductory statement at the Villach meeting was very forceful'', says Bert Bolin. ''So forceful that I myself got cold feet. Was it right to run off like this? Do we know enough to back up all of Tolba's statement?''

We did not, he answers, and he has since then put increasing emphasis on continuous critical evaluation of research results. Now, he has to safeguard the confidence that IPCC has built up by not publishing things that have not been extensively reviewed. And IPCC is under fire. A proposed update of the emission scenarios for the 1992 conference in Rio de Janeiro has come up with higher future emissions of carbon dioxide from fossil fuels, which has been criticized. Rightly, says Bert Bolin.

''I don't think this scenario is any more valid than the previous ones. I have always emphasized that we should have a spectrum of scenarios that span the whole width of uncertainty about the future. Then we can start looking into the factors that determine if we end up with higher or lower emissions.''

But he really thinks the fight about scenarios is a matter of splitting hairs. ''The matter is not one of deciding one accurate scenario of the future. We need a picture of the range of possible and plausible scenarios. Those we have show clearly that we have to start acting now, as the inertia of society is 30 to 50 years.''

The scenarios show a rise in global temperature of somewhere between 1.5 and 4.5°C. It is difficult to evaluate what that would mean, however. Regional scenarios have a different emotional and political impact. But the regional scenarios in the IPCC report are problematic and surrounded by disclaimers of all kinds as the models are not really good enough to make predictions on that scale. But why are they presented, if they do not have a sound scientific base?

''I am doubtful whether it should have been done,'' says Bert Bolin. ''But there were several people who wanted to show the work being undertaken and there is nothing wrong in summarizing results that would probably be around in the discussions anyway.

''From the beginning, only one example scenario was planned but at that suggestion several other countries wanted to be included. No one could resist and the decision had political overtones. But if the researchers do well, we should be able to explain the uncertainties.''

Bert Bolin would like to see changes in the procedures for IPCC, however. The impact assessments and the response strategies need a more stringent scientific base, similar to the one used for the scientific assessment. The same approach could be used for looking at economic consequences.

Dealing with uncertainty has been an important part of the work within IPCC. It will take until the turn of the century before we will be able to record a climate change that can be clearly associated with an increase of greenhouse gases in the atmosphere. And even when we will be able to say some things about what areas are sensitive to changes, we cannot with any accuracy say when the changes will become really important.

What people believe is, however, a completely different thing from what can be proven, says Bert Bolin. And it will become politically important how the proportion of believers to nonbelievers might shift.

"Today, I think less than 5 percent of the world population is convinced of climate warming. There is larger group of people who are worried, among them many politicians, but it is still a small percentage of the populations. Most people either think something might be in store but not during their lifetime, they don't care or have never heard of the problem.

"By the turn of the century, the first two groups might have reached 50 percent in the developed world, but not in developing countries except possibly with the attribute of yet another problem hitting them, for which the developed world is responsible. That is the political dimension which is already there," says Bert Bolin.

The political dimension of climate warming will become no less important as the discussions of who should cut down on emissions of greenhouse gases start. The IPCC is now beginning to lay the groundwork for those decisions and is at the same time feeding information into a politically very sensitive arena. The first step is to study current emissions on a national basis. When the World Resources Institute in the United States did a similar study, it received harsh criticism from an Indian environmental group and IPCC will have to tread carefully when undertaking the same endeavour in the coming two-to-three years. Each piece of information will have to be checked by people in the respective country to reach an agreement on scientific grounds before entering the political discussion.

"There are also questions of how the uptake of carbon dioxide by the seas should be credited to different nations. Equally between all people? How should one count the net uptake of carbon dioxide by some forests? Swedish forests grow more than ever and take up perhaps half of all emissions from oil and coal used in Sweden. Should we credit ourselves with that? It has been discussed, but it's clear that Sweden will not use this argument. This is the interplay between politics and science,'' says Bert Bolin.

There are also some scientific challenges ahead. When the patterns of change in climate are starting to be noticeable and may not match the model results, we have to ask why. The predicted temperature increase could differ from the observed. Why?

Part of the answer might lie in natural climate fluctuations that hide the man-made changes. But there is also another possibility. Other anthropogenic emissions. They include sulphur dioxide and the dynamics of CFCs and stratospheric ozone.

''We have to capture the total picture much better than we have done so far,'' says Bert Bolin. ''It's not as simple as a bunch of greenhouse gases acting together. We have to analyze other factors than the greenhouse gases and describe all relevant research results in a manner that is meaningful in the political discussion.''

The new information has already entered the political arena. The science adviser to the US president, Dr. Allan Bromley, has emphasized the role of sulphur dioxide and ozone in working out the US position. But is his interpretation of the new scientific findings correct, asks Bert Bolin. There is a definite risk that new scientific results surface and are used for political purposes before they have been critically evaluated.

''That is why the IPCC assessment has to be reviewed by several other researchers. Then we can say: This has been reviewed. We have to establish an awareness among politicians of not uncritically adopting the latest fad. It's better to be a year or two behind the cutting edge of research and get a more nuanced and correct picture.''

The problem of turning scientific results into political awareness is not only a matter of presenting a balanced picture, however, and Bert Bolin is rather pessimistic when it comes to political action.

"The world is not yet ready to take collective actions. We can wish it as much as we want but most people and their political representatives are not sufficiently receptive. President Bush has no mandate from the electorate. How could he act differently than he does? He could be a pioneer to stimulate discussion, which has been a successful approach in some countries in Europe including Germany, Holland and the Scandinavian countries, particularly Norway. But few countries have yet decided to keep emissions constant during the 1990s."

Could the political process have been speeded up by beginning to investigate response strategies earlier? Hardly, says Bert Bolin. With the economic system, the market system, that we all probably agree on, it is impossible for businesses to take initiative in new technologies before they sense that they can make money in doing so.

"It's a process that has to move steadily forward and I think a correct and unchallengeable presentation of the science is extremely powerful because it is something the politicians cannot argue against. And with them, a broader spectrum of the population will also be convinced."

How important is it to sign an international climate convention in 1992?

"It is definitely important because then we will not have to discuss that anymore. If we don't get it now, there will be several more years of discussion of how it should look," says Bert Bolin.

The core of the convention should be a decision to stabilize emissions in industrialized countries by the year 2000, but equally important is to establish procedures and institutions to see how the decisions are followed up. Many people will be disappointed, says Bert Bolin, and he himself does not have any far-reaching expectations.

Would it not be possible to adjust to a changing climate rather than having to change our entire energy system? It could seem like an attractive solution but hardly realistic.

"We will have to adjust anyhow, but that is not enough. The most evident example is the fate of small islands. It is much easier for a developed country to talk about adjusting than for a developing country. The Netherlands will be able to cope with a half-meter rise in sea level without having to sacrifice much in growing welfare.

"The second and most important reason is that we have to start thinking beyond the turn of the century. We know approximately how rapid the population growth will be and we know that developing nations strive toward a higher standard of living, which will require more energy. I doubt it will be possible to use less than twice the amount of energy used today in the middle of next century but use of fossil fuel will have to be less than half of the present in order to stabilize carbon dioxide concentrations in the atmosphere.

"If you quit looking only at the nearest decades and focus instead on the year 2050, you start realizing that we have to do something, regardless of any change in climate. In 2050, oil and gas will no longer be available in the way they are today and other forms of energy will be more expensive."

Discussion of adjusting has arisen, however. It is an undercurrent in some statements from the United States and the former Soviet Union has touched on that reasoning. Bert Bolin sees more of it coming. The standpoints of different countries will be dependent on their situation. So far, some countries such as those with tropical rain forest have started to group themselves, likewise the oil-producing countries. On the impact side, the African countries share a common interest with their experiences of what drought can mean.

"I hope there will not be a political fight motivated by some countries not wanting to act because they won't get hit" warns Bert Bolin. "In this perspective I almost think it's valuable that it's not possible to make reliable regional scenarios yet. This is a global problem and we have to come to agreements on how to deal with it on a global scale. There has to be some kind of solidarity."

Afterword

A look back

THE YEAR IS 2030. I AM READING A HISTORY BOOK about the decade just before the turn of the century. That was when the concern about the global environment really had entered the agenda of international politics. Scientists came with reports of ozone holes and threats of climate warming adding to the very immediate environmental problems in Eastern Europe and what used to be the Soviet Union, which was falling apart at the time. Politicians had a hard time for a while, trying to prioritize and weigh environmental concerns against each other and against other issues needing attention, not the least being poverty and the need for economic development in many poor countries. Somehow they managed, however, to reach an agreement on an international level. Also, energy saving turned out to be more economic than expected. With that as a backing and a serious discussion of how to transfer energy-efficient technologies to developing nations, a change actually started to come about. Today those discussions are thought of as a model for dealing with international problems.

Of course, the world is still far from problem-free, but at least we were able to take actions that prevented the worst of the climate-change scenarios from coming true. The sea level has risen, but not as much. It is warmer, but agriculture and forestry have been reasonably able to adjust. We are not worse off today than at the turn of the century, which I expect would have been the case if no one had been able to act at that point in time.

Further reading

● The major sources for this book has been the following scientific reports:

Climate Change. The IPCC Scientific Assessment. Intergovernmental Panel on Climate Change. World Meteorological Organization and United Nations Environment Programme. Houghton et al (ed), Cambridge University Press, 1990

1992 IPCC Supplement. Scientific Assessment of Climate Change. Intergovernmental Panel on Climate Change. Cambridge University Press.

Climate Change. The IPCC Impact Assessment. Intergovernmental Panel on Climate Change, World Meteorological Organization and United Nations Environment Programme. Tegart et al (ed). Australian Government Printing Service, 1990

Climate Change. The IPCC Response Strategies. Intergovernmental Panel on Climate Change, World Meteorological Organization and United Nations Environment Programme. 1990

The Greenhouse Effect, Climatic Change and Ecosystems. SCOPE 29. Bolin et al (ed). John Wiley & Sons, 1986

● Other useful sources:

Climate Change: Science Impact and Policy. Proceedings from the Second World Climate Conference. Jäger and Ferguson (ed). Cambridge University Press, 1991

Climate Change and World Agriculture. Parry. Earthscan Publications Ltd, 1990

Mountain World in Danger. Climate Change in the Forests and Mountains of Europe. Nilsson and Pitt. Earthscan Publications Ltd, 1991

Policy Options for Stabilizing Global Climate. Report to Congress. United States Environmental Protection Agency, 1990